Robotics:
A Manager's Guide

Rex Maus
Randall Allsup

A Wiley Press Book
John Wiley & Sons, Inc.
New York • Chichester • Brisbane • Toronto • Singapore

Computer and Technology Books for Managers from John Wiley & Sons

The Chain of Quality: Market Dominance through Product Superiority, Groocock

Expert Systems: Artificial Intelligence in Business, Harmon & King

Expert Systems Applications, Harmon

File and Data Base Management Programs for the IBM® PC, Hecht

The IBM® PC in Your Corporation, Walden

Keeping America at Work: Strategies for Employing the New Technologies, Hansen

Manufacturing: The Formidable Competitive Weapon, Skinner

Programming Expert Systems, Sawyer & Foster

Restoring the Competitive Edge: Competing through Manufacturing, Hayes & Wheelwright

Publisher: Stephen Kippur
Editor: Theron Shreve
Managing Editor: Katherine Schowalter
Production Services: The Publisher's Network

Library of Congress Cataloging in Publication Data

Maus, Rex.
  Robotics: a manager's guide.

  1. Robotics.   I. Allsup, Randall.   II. Title.
TJ211.M38   1986   629.8'92   85-29555
ISBN 0-471-84264-8
ISBN 0-471-84265-6 (pbk.)

Printed in the United States of America
86 87 10 9 8 7 6 5 4 3 2 1

To Marc, Eric, Emily, and Becky

Credits for Figures

Figure 3.2	Photo courtesy of ASEA
Figure 3.3	Photo courtesy of DEA-PRAGMA
Figure 3.4	Source: Schlesinger, G., Der Mechanische Aufbau der kunstlichen Glieden, Ersatzglieder und Arbeitshilfen, Part II, Springer, Berlin, Germany, 1919.
Figure 3.5	Photo courtesy of the National Aeronautics and Space Administration (Source: Hill, J. W., McGovern, D. E., and Sword, A. J., Study to Design and Develop Remote Manipulator Systems, final report, National Aeronautics and Space Administration contract NAS2–7507, SRI, 1974.
Figure 4.15	Source: Ottinger, L. V., Evaluating potential robot applications in a system context, Industrial Engineering, January 1982.
Figure 4.16	Source: International Standards Organization. ISO/TC, 97/SC8/WG2, Documents (N1)–(N29), International Standards Organization, 1982–1983.
Figure 5.24	Source: Hasegawa, Y., Analysis of complicated operations for robotization, SME Paper No. MS79–287, 1979.
Figure 8.1	Source: International Standards Organization. ISO/TC, 97/SC8/WG2, Documents (N1)–(N29), International Standards Organization, 1982–83.
Figure 8.4	Source: DuPont Gatelmond, C., A Survey of Flexible Manufacturing Systems, Journal of Manufacturing Systems, Vol. 1, No. 1, 1982, pp. 1–15.
Figure 9.5	Source: McCormick, E. J., and Sanders, M. S., Human Factors in Engineering and Design, McGraw-Hill, New York, 1982.
Figure 10.1	Source: Ayres, R. U. and Miller, S. M., Robotics: Applications and Social Implications, Ballinger Publishing Company, Cambridge Massachusetts, 1983.
Table 10.2	Sources: Machine-tool hours derived from American Machinist, The 12th American Machinist Inventory of Metalworking Equipment, 1976–1978, American Machinist, Vol. 122, No. 12, December 1978, pp. 133–148. Labor hours derived from Bureau of Labor Statistics (1980). Bureau of Labor Statistics, U.S. Department of Labor, Occupational Employment in Manufacturing Industries, 1977, Government Printing Office, Bulletin 2057, Washington D.C.
Table 10.3	Source: Derived from estimates of available time from statistics compiled by Ayres, R. U. and Miller, S. M., for use in Nof, Shimon, Handbook of Industrial Robotics, 1985, John Wiley & Sons Inc., New York, pp. 479.

All figures with the exception of Figures 1.2, 2.1 and 2.2 are reproduced by permission of John Wiley & Sons, Inc. from Handbook of Industrial Robotics, edited by Nof and Shimon; copyright 1985.

Table of Contents

Preface

Robots are coming of age. No longer are these mechanical marvels purely the dreams of science fiction writers. Today, they are rapidly becoming more common in the industrial world, and, as technology permits, so too will they become a part of our everyday lives in the coming decades. However, for now, robots are largely confined to the manufacturing environment, where their numbers increase every year, not just in the U.S., but world wide. Due to their successful job performance record in industry, robots are rapidly finding new applications in other diverse areas such as undersea exploration and nuclear maintenance. The recent remarkable exploration efforts in locating the wreckage of the luxury liner *Titanic* were made possible by unmanned robot "divers." The meticulous clean-up and inspection of the Three Mile Island nuclear reactor were also made possible by using robots. The potential uses for these machines is only as limited as technology itself. They are here to help us become more efficient in jobs we find distasteful or dangerous. They are here to stay.

We have been fascinated with robotics for some time, and, by our extensive CAD/CAM/CAE (computer-aided design/computer-aided manufacturing/computer-aided engineering) backgrounds, the study of robotics became a natural extension for which to apply our efforts. We are consultants by trade, specializing in technical documentation and marketing communications for the computer field. We have worked on different CAD/CAM/CAE projects for some of the largest companies in America. As a result, the more we learned about CAD/CAE, the more we became interested in CAM and robotics, the primary reason being that the manufacturing environment is where CAD/CAE disciplines become tangible money-making entities. Manufacturing is also an area that is relatively untouched by computer technology. So, the more we learned about CAD/CAM/CAE, the more we were convinced that robotics would change manufacturing as we know it today. Robotics will continue to revolutionize the ways in which industry and its management think and act.

We have read and studied robotics extensively and have held discussions with many people representing manufacturing companies. From research and discussions, it soon became obvious that the new technology could offer significant benefits that could help American industry regain its competitive edge. America's major industrial competitors were using the technology against us profitably, so we felt that U.S. industry should have the same advantage. As a result, we developed this book to serve as a primer for manufacturing management. It is intended to illustrate the capacity robots have to reshape the American manufacturing environment.

We have written this book for the executives, middle managers, and computer systems personnel of industry. These are the people who most need to know how robotics can impact the manufacturing workplace. Rather than trying to encompass everything about robotics and CAM, we have instead included information that will provide a sound base of knowledge for any person interested in the field. You will not find a lot of technical jargon in the book. We have tried to relate the principles of robotics, as well as robotic applications themselves, in a logical manner that directly reflects the world of manufacturing. Where advanced concepts are addressed, they are explained in simple terms. We intend this book to be used as an introduction to the world of robots, not as a technical reference (there are already plenty of these books).

To make it easier for you to use the information included in the book, the following paragraphs briefly explain what is included in each chapter.

Chapter 1 provides an overall introduction and focuses on the current state of American industry. It should not be construed as a "gloom and doom" type of dissertation, as it is intended to establish how we have lost our position in world markets due to a complacent attitude of industry. It also introduces the rationale behind automating the factory.

In Chapter 2, we discuss the manufacturing environment, breaking down industry into the sectors of design, fabrication, assembly, inspection, and materials handling and discussing problems found in these sectors today. Finally, we discuss current management trends in industry and introduce the concept of the technological manufacturing manager.

Computer automation is introduced in Chapter 3. Here, we compare the introduction of computer technology into the business world with the current introduction of robotics into the world of industry. We use several examples to illustrate how computer automation has helped business become more efficient. We also use this chapter to discuss the Japanese and their uses and attitudes toward robotics. This discussion provides a sound point of reference, as the Japanese are the most innovative makers, as well as users, of robots today.

Actual uses and applications for robots are discussed in Chapters 4 through 8. Different types of robots are explained, as are their uses in real-world applications. Each chapter contains robotic information that pertains to the different manufacturing sectors—design, fabrication, assembly, inspection, and materials handling. There are numerous illustrations in these chapters that show robots being used in all types of manufacturing tasks.

Chapter 9 pulls the entire picture together with its discussion of flexible manufacturing systems (FMS). FMS is what American industry is striving for; automating the entire factory, from the paper shuffling of inventory and materials management to the machine center scheduler to the processes of automatic tool changing. The concept of computer-integrated manufacturing (CIM) and its future is covered in the last part of the chapter.

Chapter 10 introduces the real reasons for using robots in manufacturing—economics. Efficiency, capacity, and capital costs are discussed in nontechnical terms. You do not need to have an economics background to use the information in this chapter. The chapter provides the manufacturing manager with background information in order to make further consideration possible and worthwhile.

The factory of the future is discussed in Chapter 11. It is an educated guess as to future uses and trends in the field of robots and their expanded uses in industry. But, just as computer automation has expanded tenfold in the past decade (far surpassing even the best prognostications by many experts), robotics technology has the same potential for expansion. This chapter discusses possible further robot enhancements, as well as introducing applications for the next generation of robots, where artificial intelligence is expected to play an increasing role.

The appendixes contain additional information. Suggested readings are given in Appendix A; they will be useful to supplement your further study of this exciting technology. Appendix B contains a glossary of robotic terms used in this book. Appendix C presents tables and charts that illustrate robot installations, uses, and the vendor market share worldwide. Finally, robotic organizations and robotic vendors around the world are listed in Appendix D. To make this book more concise, which is consistent with its design as a primer, we have avoided the use of extensive footnotes and references.

We would particularly like to thank Shimon Y. Nof, editor of the *Handbook of Industrial Robotics* (a Wiley book). This book is the definitive technical reference for robotic technology today. It is without peer. Many times we found ourselves in need of precise technical information regarding a particular area of robotics, and Mr. Nof's book always provided us with the answers. Without his handbook, our book would not have been possible.

Many other people assisted us in the research and preparation of this work. Without their knowledge and advice, it would never have become a reality. To all of you, many thanks. We'd especially like to thank Jerry T. Maxwell—our benevolent, straight-shooting attorney, Theron Shreve—a friend and confidant, and all of the many other helpful people at John Wiley & Sons for their assistance and patience. Finally, of course, a hearty Texas thank you to our friends and families for their continued support throughout the project.

Rex Maus
Randall Allsup

Introduction

Since the industrial revolution, advances in technology, no matter how slight, have helped the United States grow into a position of manufacturing leadership. This leadership made us economically successful and put us in the position of being a respected trading partner with other countries. We produced the finest goods that were available anywhere in the world. In fact, the world came to us for just about everything it needed.

Manufacturing Today—A Bleak Portrait

Our manufacturing industries have lost their competitive edge. The economy is suffering, and our indebtedness to other nations grows daily. Once again, we are turning to technology to provide the impetus to revitalize our lagging industrial base and give us renewed prominence in world markets. Hopefully, the new technology in manufacturing automation can put America back on track. What is this new technology and how does it work? Can it really be counted on to get the American manufacturing industry back on its feet? And, if so, what will be the costs, both economically and culturally?

This book examines computer automation and the robotic revolution and how the impact of these new technologies will affect the future of the American workplace. The concepts and techniques described in this book will revolutionize how American industry operates in the next 15 years. The factory of the future will begin to emerge, where machines take the place of humans and with products becoming less expensive to produce while attaining higher quality. The current manufacturing methods will be swept away in the same fashion that computers have replaced most of the manual office tasks during the past 10 years.

America became industrialized during the nineteenth century. Newly developed machines greatly expanded the productivity of American workers and factories. Manual methods of production were slowly replaced by simple, but productive, machines. New inventions were almost commonplace and caused the United States to become a recognized leader in manufacturing expertise. Examples of our technological leadership abound when you consider the development and implementation of steel mills, machining advancements, automated looms, transportation techniques, and other significant discoveries and inventions. Figure 1.1 chronicles many of the advancements and

Inventions and Discoveries of the United States

1780-1810	1810-1840	1840-1870
Cotton gin Whitney (1793)	**Gun breechloader** Thornton (1811)	**Lathe, turret** Fitch (1845)
Plow, cast iron Newbold (1797)	**Magnet, electro** Henry (1828)	**Printing press, rotary** Hoe (1846)
Propeller screw Stevens (1804)	**Locomotive, first U.S.** Cooper, P. (1830)	**Sewing machine** Howe (1846)
Steamboat Fulton (1807)	**Flanged rail** Stevens (1831)	**Turbine, hydraulic** Francis (1849)
Paper Machine Dickinson (1809)	**Mowing machine** Manning (1831)	**Ice-making machine** Gorrie (1851)
	Reaper McCormick (1834)	**Locomotive, electric** Vail (1851)
	Pistol (revolver) Colt (1835)	**Elevator brake** Otis (1852)
	Telegraph Morse (1837)	**Mason jar** Mason, J. (1858)
	Babbitt-metal Babbit (1839)	**Sleeping car** Pullman (1858)
	Rubber, volcanized Goodyear (1839)	**Shoe-sewing machine** McKay (1860)
		Machine gun Gatling (1861)
		Printing press, web Bullock (1865)
		Refrigerator car David (1868)

Figure 1.1

Inventions and Discoveries of the United States

1870-1900	1900-1930	1930-1960
Compressed air rock drill Ingersoll (1871)	**Steel alloy, high speed** Taylor (1901)	**Refrigerants** Midgley & coworkers (1930)
Engine, gasoline Brayton (1872)	**Airplane with motor** Wright brothers (1903)	**Nylon** Dupont labs (1937)
Car coupler Janney (1873)	**Bottle machine** Owens (1903)	**Computer, automatic sequence** Aiken, et al. (1939)
Telegraph, quadruplex Edison (1874)	**Radio amplifiers** De Forest (1907)	**Helicopter** Sikorsky (1939)
Gun, magazine Hotchkiss (1875)	**Washer, electric** Hurley Co. (1907)	**Transistor** Shockely, Brattain, Bardeen (1947)
Telephone Bell (1876)	**Air conditioning** Carrier (1911)	
Phonograph Edison (1877)	**Gyrocompass** Sperry (1911)	
Welding, electric Thomson (1877)	**Lamp, klieg** Klieg, A & J (1911)	
Cultivator Mallon (1878)	**Lamp, mercury vapor** Hewitt (1912)	
Lamp, arc Bursh (1879)	**Wrench, double acting** Owen (1913)	
Lamp, incandescent Edison (1879)	**Dynamp, hydrogen cooled** Schuler (1915)	
Punch card accounting Hollerith (1884)	**Telephone, radio** AT & T (1915)	
Trolley car, electric Van DePoele (1884)	**Gun, Browning** Browning (1916)	
Monotype Lanston (1887)	**Radio crystal oscillator** Donovan (1918)	
Motor, induction Telsa (1887)	**Spectroscope** Dempster (1918)	
Camera, Kodak Eastman, Walker (1888)	**Elevator, push button** Larson (1922)	
Harvester - thresher Matteson (1888)	**Gas discharge tube** Hull (1922)	
Meter, induction Shallenberger (1888)	**Gasoline, leaded** Hidgley (1922)	
Automobile, steam Roper (1889)	**Radar** Taylor, Young (1922)	
Pneumatic hammer King (1890)	**Arc tube** Alexanderson (1923)	
Steel, alloy Harvey (1891)	**Welding, atomic hydrogen** Palmer (1924)	
Submarine Holland (1891)	**Circuit breaker** Hilliard (1925)	
Automobile, electric Morrison (1892)	**Engine, gas, compound** Eickemeyer (1926)	
Automobile, gasoline Duryea (1892)	**Television, electric** Farnsworth (1927)	
Motor, A.C. Telsa (1892)	**Teletype** Morkrum, Klienschmidt (1928)	
Plow, disc Hardy (1896)	**Coaxial cable system** Van de Graff (1929)	
Stove, electric Hadaway (1896)	**Rocket engine** Goddard (1929)	
Turbine, steam Curtis (1896)		
Turbine, gas Curtis (1899)		

Figure 1.1 (continued)

discoveries that were made during the nineteenth and twentieth centuries.

It is interesting to note that the most recent significant dicovery listed in Figure 1.1, the discovery of the transistor, occurred in 1947. True, other significant discoveries have occurred since, but there have been no recent inventions or discoveries that are anywhere near the magnitude of those from the nineteenth and early twentieth centuries. However, all of these discoveries had one thing in common; they all required an ability to produce and manufacture the products. This is the main reason American commerce grew so astoundingly.

All through the nineteenth century, America industrialized her factories, implementing new machines that expanded productivity and raised economic prospects. Many individuals became very wealthy simply because their discoveries eliminated old manual methods for producing goods. America possessed the world's leading manufacturing companies. With the twentieth century came more advances in technology, including machines capable of mass production and the introduction of Henry Ford's assembly line. America outdistanced all others in both productivity and prosperity. We were the example the world looked to as a model of industrial efficiency. If it said ''Made in the U.S.A.'' it had to be good.

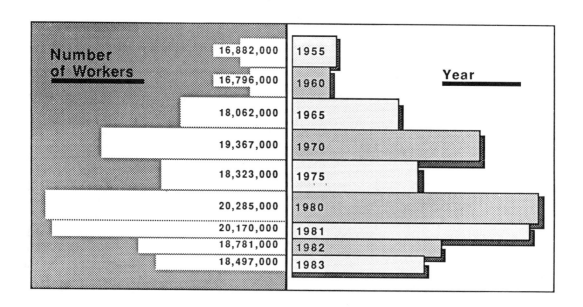

Figure 1.2 Employees in manufacturing jobs (Source: Bureau of Labor Statistics, U.S. Labor Department)

Today, in spite of our past accomplishments, many sectors of the American manufacturing community are bankrupt or soon will be. In the last decade, over two million manufacturing jobs have been lost, while more than 1,000 companies have closed their doors for economic reasons. Ample evidence of the decline of American manufacturing is provided by the statistics in Figure 1.2 regarding manufacturing employment. These statistics document the loss of workers that has resulted from the declining status of U.S. manufacturing in world trade. The phrase "Made in the U.S.A." has little if any meaning, except to Americans. The outlook is indeed bleak. Where we once produced over half of the world's steel, we now import almost 35 percent just to meet our own needs. Our steel industries try to weather the storm, but they are doing little to counter its effects.

Where then, does the blame for this incredible downturn lie? Some manufacturing gurus say that it is the fault of management, others say it is because we have not replaced antiquated manufacturing equipment, while still others say it is because of labor unions. Some people claim the fault lies in the methods we use in manufacturing, and others feel that it is because of the government of the United States itself. The truth is that it is the fault of everybody, from the government and the top management of our industries down to the foreman of a six-person metal fabrication shop.

We have stagnated, and our intuitive renaissance attitude has been replaced with sheer complacency. With complacency comes fewer innovations and less input for new advancements to be made. It is a simple truth—you can't build a world-class economy without having world-class methods of production. This is readily seen if you examine what is being put into American industry. The input includes the basic skills of the work force, the quality of the managers, the capital investment into industry, our research and development methods, and our current applications of industrial production. Each input today shows its own degree of complacency.

Leadership and ingenuity are missing and have been for decades. It has been said that if the U.S automobile industry had responded with the same actions as the American computer industry has in the last five years, that domestic automobiles would be lighter in weight, average over 50 miles to the gallon, and sell for less than $1,000. Input is very important. By thoroughly examining manufacturing inputs, and making comparisons with other segments of our economy, it is easy to see that American industry does not stack up favorably, especially when comparisons are made with our major foreign competitors.

Foreign competition has become a sore spot that constantly antagonizes our national leaders, manufacturing experts, and general population. It is because of foreign competition (as well as our own laissez-faire attitudes) that our balance-of-trade deficit continues to grow. The U.S. now imports 25 percent of her cars, 40 percent of her vacuum cleaners, 100 percent of her video cassette recorders, and almost 100 percent of her clothing and footwear. Table 1.1 illustrates the dismal U.S. trade picture by contrasting our import and export levels.

Table 1.1 United States export/import statistics

U.S. Exports and Imports of Leading Commodities (in millions of dollars)

Source: Office of Industry and Trade Information, U.S. Commerce Department

Commodity	Exports		Imports	
	1982	1983	1982	1983
Ores & Metal Scrap	2,174	2,276	2,684	2,500
Lumber & Rough Wood	2,095	2,104	1,737	2,719
Coal	6,072	4,115	---	---
Petroleum Products	5,947	4,557	59,396	52,325
Natural Gas	---	---	5,934	5,530
Office Machines & Computers	10,206	11,669	4,929	6,759
Motor Vehicles & Parts	13,907	14,463	25,246	46,247
Tires & Tubes	372	304	1,239	1,406
Agricultural Machinery	2,389	1,589	1,042	1,196
Paper & Manufacturers	2,654	2,553	3,848	4,215
Iron & Steel Mill Products	2,101	1,415	9,184	6,338
Nonferrous Base Metals	1,768	1,606	5,321	7,422
Clothing	952	818	8,165	9,583
Footwear	---	---	3,438	4,010
Toys, Games, Sporting Goods	957	878	2,786	2,506

We are rapidly becoming totally reliant upon trade imports, reversing the trends of just 25 years ago. Part of our problem is that we do not recognize that there are manufacturing problems in our country. As discussed by John Naisbitt in his book *Megatrends*, our devotion to the American way of manufacturing is causing us to ignore the fact that we are being beaten at the very game we created. We ignore the facts themselves, falsely assuming that the industrial techniques that got us this far could not let us down. As Naisbitt says, "Its [American industry's] demise was—for many of us—unthinkable."

When we do admit that the rest of the world is beating us, we say they are only doing so by cheating. Foreign governments are stacking the deck against us by providing subsidies and incentives for their own industrialization. But really, what is wrong with this? It illustrates a progressive spirit that is missing in America, the birthplace of free enterprise. Also, many manufacturing managers, including some of the top executives in this country, say that the Japanese are unfair. This, too, is totally ridiculous. What is unfair about a country producing a cheaper, higher-quality product? This is what free enterprise is all about. We certainly were not complaining when other nations stood in line waiting for the opportunity to buy our products 50 years ago.

The U.S. economy is in transition, and we are gradually shifting to an economy that is structured around service work rather than manufacturing. This parallels the trends representing the shift toward the information age. Just as America shifted from an agricultural society to a manufacturing society, we are now moving from manufacturing to information. However, an information-based society will only help manufacturing, not hinder it. It will help provide the new start that is badly needed within our factories.

Change is in the air. As in the nineteenth and early twentieth centuries, revolutionary advances are on the way, heralding a promise of rejuvenation in the American manufacturing community. They are the forces of computer-driven automation. The forces range from computer-aided design and computer-integrated manufacturing to computer-controlled programmable robots. Industry is beginning to use the tools of the information age— using computers to manipulate information and to control machines, using machines to regain our competitive edge.

America is entering a major era of reindustrialization that will draw upon the incredible power of current and future computer technology. Computer technology will alter manufacturing in this country. This new era will include the implementation of new, more productive machinery to replace antiquated equipment. It will provide for the redesigning of the majority of our products for automated production methods, change the attitudes of manufacturing management, and reshape the lives of the American industrial work force.

Computers, robots, and other tools of automation can play a vital role in revitalizing our manufacturing methods. Automation can be a central force in boosting our economy if it is used in beneficial ways. None of this is going to happen if American manufacturers continue to think that technological change will bring about chaos and disruption rather than progress and productivity. The future of manufacturing is not something that we can worry about in a few years. To become competitive at home and abroad, the future is now.

Are computer automation and robotics really the keys to revitalizing the American manufacturing system? If so, how will the United States pursue and implement these revolutionary new tools? How will computers and robots provide the means for American industry to be revitalized? How will we use automated robots to produce better-quality goods at lower costs? What benefits can computer and robotics technology deliver to the American manufacturing industry? Is greater productivity really possible by automating our industries? Can we meet and conquer foreign competition simply by producing goods of better quality at lower cost? Is manufacturing flexibility merely a statement, or can it really be achieved? Can we reduce our labor costs substantially in our manufacturing endeavors? How much could computer automation and robotics change and disrupt our traditional industrial methods?

These questions are typical of top management in American industry today. Obviously, change is inevitable; you witness change every day of your life. Society and industry are subject to change also. American manufacturing is changing; it has to. This book discusses these changes. It examines changes in U.S. manufacturing that are made possible through the use of the new technologies of computer automation and robotics. It introduces and explains the concepts of robotics and computer automation and describes what automation is capable of doing. It discusses the different automated systems and explains how these systems might be developed so that the greatest benefit, socially and economically, can be attained. It describes computer automation and robotics applications that are used in the manufacturing cycle, and it provides insight regarding future technologies.

We have an unprecedented array of choices before us regarding the implementation of computer automation and robotics in our factories, because automation and robotics affect a wide variety of manufacturing applications. We also have the unprecedented potential to use this computerized automation in ways that benefit not only industry, but society as a whole. Automation in manufacturing is not isolated to unique factories or specialized application; its effects are far-reaching and revolutionary. Just as our nation has grasped previous opportunities that affected and enhanced industry, it would be tragic not to seize the potential for productivity and economic success today and use this new technology in ways that will bring real tangible benefits to our factories as well as our individual lives.

The Outlook

As recently as the mid 1970s, computer-automation and robotics applications were largely confined to university and corporate research laboratories. Today, major computer-automation and robotics programs are being launched in many Fortune 500 companies. Granted, the current applications involve "showcase" factories of large corporations, such as IBM, General Electric, General Motors, Chrysler, and John Deere, but others are following as the trail is blazed. Many Fortune 500 companies are implementing robotics automation, as progressive managers weigh the risks involved and forge ahead in search of increased productivity and profits.

Robotics systems will change the way industry operates by altering the way that we manufacture goods. This new technology will make it possible to develop a wider variety of goods and products with consistently higher quality, lower production costs, and the capability to make product changes efficiently and cost-effectively. Robotics and computer-automation will also help America solve its productivity problems. New technology will help industry reorganize itself into more efficient and effective manufacturing organizations by providing industry with the capability to solve manufacturing problems more quickly and efficiently. Factory managers and workers will have powerful computerized workstations that they can use to handle critical information pertaining to the operations on the factory floor. Bottlenecks and disruptions might be virtually eliminated. Instead of waiting for production reports, computer-automation will provide managers instant access to productivity information in all sectors of the factory environment.

Using computers, factory managers will be able to monitor all factory activities and personnel while using the computer to increase the quality and quantity of their manufacturing decisions. Computers will monitor all factory equipment, scheduling, problems, and other items related to the factory environment. In short, the entire manufacturing process will become much more rational. Manufacturing information will be gathered, manipulated, and put into useful forms that will directly increase productivity and efficiency.

Consider, for example, the problems of design and fabrication in manufacturing a machine part used in the assembly of the space shuttle. The specifications of the part change frequently to meet the needs of a particular space mission. You can do design work and tool up to produce the original part at nominal expense, but as the requirements for a new part are changed, the redesigning and retooling required to produce a one-of-a-kind part becomes very expensive, and the time involved for modifications is lengthy. Imagine instead that you could use computer-automation and robotics technology to eliminate the long and expensive design and retooling processes, so that you could produce a variety of custom parts quickly and efficiently.

It almost sounds too good to be true, doesn't it? This sort of activity is a reality. Today, manufacturing concerns that use computer-aided design (CAD) systems and robotic machining applications perform this type of activity every day. The CAD system eliminates the problems of tedious redesign work, while the reprogrammability of the robot machining tools makes fabricating a new part an almost effortless process. This type of process is impossible using today's traditional manual methods of manufacturing.

The introduction of computer-automation and robotics technology in our factories will prove exciting to manufacturing experts and professionals. The new technology will help manufacturers define problems and determine what automated processes need to be implemented to solve manufacturing problems in ways they have never considered before, simply because manual methods offered little, if any, alternatives. And, as more factories automate, corporate decision makers will be free to focus on the more difficult aspects of manufacturing. This will result in even more solutions, and perhaps even more automation, in the attempt to eliminate manufacturing problems. As problems are eliminated, productivity increases, as do profits. As the old saying goes, ''Build a better mousetrap, and the world will beat a path to your door.''

Background

In reality, robotics is a peripheral extension of a computer, where the robot, of which there are many different types, receives all of its instructions from a central computer. Naturally, the efficiency of the computer dictates the efficiency of the robot. In this book, however, robotics technology is the primary area of discussion.

Long before they were used in industry, patents for robots had been filed with the U.S. government. Robot research actually began in the early 1950s, with the developmental work of George C. Devol, the father of robotics, on a programmable manipulator. The manipulator was a strange-looking device, a large pedestal with an attached protruding metal arm, vaguely resembling a human arm, connected to a pseudo-wrist with a two-piece angular pincher that acted as fingers. The manipulator had limited degrees of movement, basically involving the arm itself, which could swivel, combined with primitive yaw, pitch, and roll movements of the wrist. The "fingers" could open and close, which enabled the arm to grasp and release an object. It was primarily designed for simple pick-and-place, fixed sequence, applications. In 1956, Devol applied for the patent for his marvel.

In 1961, patent number 2,988,237 was issued to Devol for his manipulator. Experts generally agree that this was the birthdate of robots in this country. At the most basic level, Devol's robot was simply a programmable machine designed to perform pick-and-place tasks that were originally intended to be done by humans. Although Devol's manipulator was simple, it worked. Today, some sophisticated robots can work even better than humans, because they perform tasks repeatedly with only small deviations in accuracy. However, the attraction for early robot usage centered on the fact that a robot could operate in an environment that a human could not withstand. Devol understood this. This is the essence of robotics technology: machines that are capable of performing a given task better, faster, safer, and cheaper than a human being can. Devol correctly assumed that robots would be the next step in our continuing industrial revolution.

After Devol received his patent, many manufacturing managers regarded his discovery as a science fiction oddity, and they refused to believe that this primitive contraption could do all of the things he claimed it could. Devol had a supporter, however—Joseph Engelberger. Engelberger, a pioneer in the area of robot manufacturing, thought that Devol's concepts had true merit. They shared a dream and began to work together.

In 1956, Engelberger founded Unimation, Incorporated, America's first robot manufacturing firm. These two men had the foresight to envision robots being used to perform a variety of manufacturing applications. They saw programmable robots performing tasks that were too dangerous or tedious for humans to do. Together they conducted the first robotics applications study beginning in 1956, regarding the use of robots in factory automation. They visited automobile assembly plants and other manufacturing operations to gather data to determine where robots might replace manual assembly tasks being performed by humans.

They quickly surmised that robots could be used on a wide scale to do the most dangerous and repetitive jobs in the manufacturing environment. The jobs themselves were often simple, but they required exposing people to dangers in the environment, mundane repetitive acts, or the lifting and moving of heavy objects. Robots were perfect for these applications, and both men knew it.

From their studies, they began Unimation's first robot development project, designing and building a prototype robot, based on Devol's first manipulator, that could be used in the dangerous application of loading and unloading a die-casting machine. The project had its difficulties; the main problem was designing the electronics that controlled the manipulator's functions. In the late 1950s, vacuum tubes were the order of the day, as solid-state electronic components were not in wide use, although they were available. The engineers at Unimation designed a functioning electronic controller that was very cost-effective and ahead of its time. This was a remarkable achievement when considering the fact that a five-axis controller for a numerically controlled machine tool (using paper tape technology) could be obtained for about $30,000, a large investment in the 1950s. Unimation produced a successful, working design for under $10,000, a truly significant milestone. The actual cost breakdown was 75 percent for electronics and 25 percent for the mechanical parts. Today, this cost ratio is reversed. By 1961, the prototype had been thoroughly tested and was ready to be implemented in a real-world manufacturing environment. That year the robot was installed and began work, loading and unloading a die-cast machine in a General Motors fabrication plant. This robot worked efficiently and productively for years until it was replaced with a newer model. It was retired to the Smithsonian Institution in Washington, D.C. This robot is shown in Figure 1.3.

As you might expect, there were soon many other individuals and companies that jumped on the bandwagon. Unfortunately, they quickly realized that this was not an inexpensive industry to survive in, and many soon went under for lack of financial support. Incredible as it may seem, Unimation, after developing and installing what many regard as the first automated robot for factory use, continued to fight traditional manufacturing methods and did not show a profit until 1975. Some of the others did survive, but they operated on shoestring budgets with incredibly small staffs. Between 1961 and 1975, robotics expanded, and many new robots were developed and put to work in the American factories that had the most money, the most need, and savvy management. There are still some early robots on the job today, having already logged well over 100,000 man-hours; that is over 50 man-years of labor, in continued and steady production. Almost in spite of itself, robotics continued to expand although very slowly, and at high costs. By 1970, there

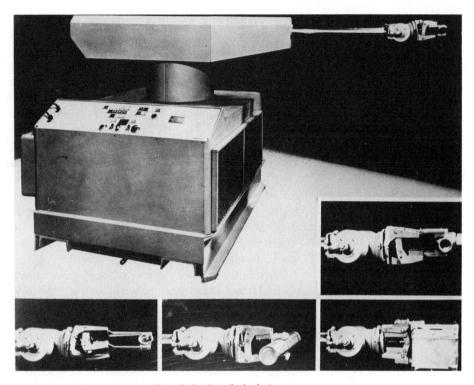

Figure 1.3 Engelberger and Devol's first installed robot

were enough manufacturers of robots and robotics technology that a symposium was held in Chicago, drawing a meager attendance of a little over 100. The concept was expanding, however, and new ideas and applications research were creating more enthusiasim for the growing field.

In 1975, the Robot Institute of America was founded, giving robot developers, users, and manufacturers a vehicle to share ideas and pitfalls. The true turning point in robotic evolution came in the late 1970s, when computer automation crossed the cost-prohibitive barrier; the microchip made computers faster, cheaper, and more powerful, and robotics was to benefit directly. Prior to this development, the limiting factor in robotics applications was the limits in computer processing power and programmability. The new generation of computers provided the necessary memory requirements, processing speed, and, most important, an affordable price. By 1979, robotic development was moving ahead in research and development, as well as in the installed base of working robots. People's attitudes began to change about automated equipment. The computer was becoming part of our lives, and robots no longer seemed so odd.

They had left the world of science fiction and entered the real world of industry and science. The second generation of robots was being designed.

Today, robotics research continues in the United States, Europe, and Japan, and the number of manufacturing firms implementing computer-automated robots is increasing world wide. Robotics implementation and research are expensive, but the gains are well worth the costs. Robots are becoming more sophisticated and are capable of performing a wider variety of complex procedures than ever before. Currently, there are almost 100,000 robots at work in the world, with Japan and the United States using 69 percent of the total. American factories employ only about 15,000 robots, displacing 20,000 human workers, but there are still well over 19 million factory workers performing mostly antiquated, inefficient manufacturing tasks in American industries.

As you can see, the events that are commonly called the industrial revolution are still occurring. The revolution is not over but has entered a new phase called the information age. The computers that have been storming the country for the last several years can now move and act on their environment through robots. This is merely the next step in the evolution of manufacturing automation. The age of robotic automation has begun.

There is much more to the world of robots than can be covered in a single text. Robots have applications other than in the factory, but the factory has been the real mover in the development of robots. Undersea exploration and agriculture stand to benefit greatly from robots, but how much can those fields further robotics advances? Space exploration also holds vast possible uses for robotic technology, but here, too, applications are limited by environment. History seems to back the claim that the manufacturing sector will be the greatest market for new technology. One of the reasons for the preponderant role of industry in the advances of new technology is that industry has benefited greatly from new technology in the past and is likely to do so again. Another reason is that industry is already set up for machinery, and manufacturing is a scientific discipline, engineered with great care and planning. Plant engineers are used to thinking in quantified terms, carefully planning each stage of the production process. These qualities are exactly what is required to successfully implement robotic systems.

The same computer installations that have aided the design and manufacture of products can provide additional impetus to automated materials handling and other process-type production operations. Greater data-handling abilities have enabled computers to control more sophisticated processes, going hand in hand with advances in servocontrol motors and improved vision systems. Numerical control programs for milling machines can be generated on desktop computers, tested in the lab on model apparatus, and sent to a flexible robot workcell, which completes all phases of production.

The coming years will bring an increase in the development and use of the mechanical marvels. Companies are attempting to stake a claim in this exciting industry, hoping to reap high profits from the sales of their computer-automated machines. In the coming 10 years, many of the current robotic companies will collapse, and others will totally change the products that they make. The survivors will develop more complex robots that are capable of tasks we can only dream about now. These machines will change the world of manufacturing as we know it today. They will change society itself. Hopefully, they will enable the American manufacturing community to reemerge as a dominant force in the global trade picture. It will require major investment, major commitment, and a new way of thinking, but when considering the results that are possible, can we afford not to continue to develop and use these tools of automation?

The Manufacturing Environment

This chapter focuses on the methods used in manufacturing today. It examines the factory environment and discusses the different areas of manufacturing, including design, fabrication, assembly, inspection, and materials handling. This chapter is a prelude to the discussions of robotic automation in the factory environment in Chapter 3, and it also discusses the major problems that affect each of the sectors of manufacturing, and finally, it covers management trends in the current workplace.

Sectors of Manufacturing

Manufacturing is a complex and intertwined activity. In order to discuss the manufacturing process, it is necessary to break manufacturing down into its smaller, more manageable parts, the islands of applications. The islands of applications include design, fabrication, assembly, testing, and materials handling. Figure 2.1 illustrates the islands of applications. The islands themselves have traditionally been segregated in American industry, as this provides managers with the opportunity to exercise more control over each sector. However, by segregating applications, there may be more management control within each sector, but there is less integration between sectors. This causes many problems in the manufacturing cycle. The integration of manufacturing sectors has been studied for some time by leading corporations and universities, producing few results that would significantly improve current methodology in manual operations. It will be the computer that revolutionizes and integrates the sectors of manufacturing.

Design

The design sector is where an idea becomes a concept. Before any product or part can be manufactured, it must first be designed. The design process involves taking the basic idea for a part or product and translating the idea into a picture. In the traditional manufacturing environment, design is a manual task performed by engineers, architects, and drafters. It is extremely time-intensive, because it requires meticulous attention to detail. In the design stage, a product is usually broken down into its unique components, and each component is drawn by hand

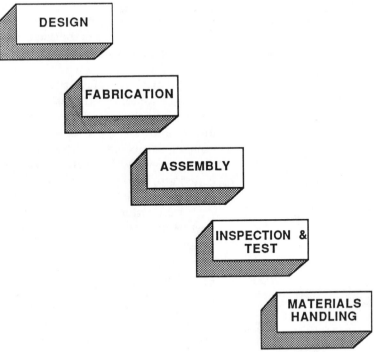

Figure 2.1 The sectors of manufacturing

using traditional drafting tools, T squares, triangles, rulers, protractors, compasses, pencils, pens, paper, and various templates. A completed design instructs the fabricators how to build the design step by step.

The problem with manual design work is that it is paper-bound; changes are often just as difficult as originals, and the process may take hours or weeks (even months), depending upon the complexity of the drawing and the number of layers it contains. Once a drawing is made by hand, it is mechanically translated into a blueline rendering. From the bluelines come the actual blueprints that are used in the fabrication process.

Because each drawing is a unique piece of artwork, changes or revisions go through the same time-consuming process as the original drawing. Imagine that after spending weeks designing an assembly for an automobile frame, minor changes are needed to produce a more rigid, sturdier assembly. In order to make the necessary modification, the design engineer or drafter must redraw the design by hand, for each affected layer, to make the required minor change. This type of activity is very cost-ineffective and wastes man-hours and profits. Often, designs that look perfect on the drawing board cannot be made on the machine floor. Obviously, this leads to further design work. However, without design drawings, there would be no products or parts; they are that important.

Fabrication

The next sector in the manufacturing process is fabrication. In this island of manufacturing, design drawings take a trip to the floor of the factory, where they are developed using real materials and machines. Here, the instructions provided by the design drawings are followed by machinists and craftspeople to make a real part or product. For example, the 5,000 components of an automobile are each fabricated (by different companies, of course), inspected, and tested before they are shipped to the assembly plant. The early development of a design is usually called a prototype. There may be numerous prototypes developed before the final product is ready to be machined.

Fabrication involves taking the hand-drawn designs for a part or product and having the machinist or craftsperson make the part or product by following the blueprints of the design. This process is also very time-consuming, depending upon the complexity of the design.

After a successful prototype is fabricated and tested for accuracy, stress, or tolerance, the part or product is ready to be produced in quantity. If, when the part is tested and inspected, a fault is found, the part is either refabricated and again retested and inspected, or it is returned to the design sector for further work. The fabrication sector can become very difficult when the time comes to begin the production of a new or revised product or design. In traditional manufacturing when a factory wanted to make a change, it required the manufacturer to change the machine or tooling cycle that was involved in fabricating the product. This is an entirely mechanical process that is accomplished either by changing fixtures or tools. It is a time-consuming process that eliminates productive fabrication time. Time and money is lost when this method is used.

Once the final prototype is complete, it is ready for the next fabrication step. This fabrication process involves the production, either in volume or as a one-of-a-kind item, of the final prototype fabricated version. Most often, fabrication involves the machining (cutting, drilling, shaping, forging, molding, etc.) of the part or product by humans, using manufacturing equipment. There are innumerable machines that have been used in the fabrication process for decades, providing manufacturing with an automated process for the production of just about any product you can imagine. It is interesting to note, that about 30 years ago when production automation began using different machines, the instructions that told the machines how to operate were stored on paper tape that had been encoded by holes punched in the tape. The tape resembled the music roll that is used with a player piano to instruct the piano what keys to press to make a tune. With these automated machines, a machinist or craftsperson is responsible for taking the design data from the blueprints and making the punched tape that instructs the machining tool how far to cut, how deep to drill, etc. The holes in the tape basically provide geometric coordinate information for the positioning of the component to be machined, as well as information that governs the operation of the machining tool itself. This primitive automated process enables machinists to use their extensive knowledge of the machining processes, tools, and experience to translate design information into machine commands to produce the desired part or product.

Parts or products that must be put together then proceed to the next manufacturing sector, assembly. Parts designated for assembly are usually checked after fabrication to make sure they were machined to the proper specifications. Those parts or products that do not meet the design requirements are rejected. Parts or products that do not require assembly are inspected and usually readied for shipping.

Assembly

The assembly sector of manufacturing is where parts that have been fabricated are put together, again following the design drawings, to produce a complete component or part. Assembly is a labor-intensive process where humans perform most of the work by hand. Some assembly tasks require only simple joining of parts with brads, screws, or other fasteners, while other assembly functions are more complex and require labor-intensive operations such as welding. Sometimes they require specialized machines or other equipment. The assembly process is usually the final process before inspection.

Materials Handling

This sector of manufacturing involves the movement of raw materials, transport of workpieces being fabricated between machining centers, and the conveyance of finished parts or products to storage areas in the factory. Materials handling deals with delivering all of the necessary parts, raw materials, and workpieces to the right place in the factory at the precise time when the parts or supplies are needed.

Materials handling is a task almost entirely performed by humans using mechanized carts, trucks, forklifts, and other means for transporting materials. Although it does not seem as important as the other sectors of manufacturing, materials handling has the responsibility of getting the proper materials where they need to be. For example, if the materials handler responsible for getting raw materials to the factory floor is negligent, the fabrication sector is essentially shut down for lack of workpieces. On the other hand, if materials are allowed to stack up after assembly and inspection, this creates a bottleneck, causing production to cease temporarily, while the finished goods are moved out of the way of production and conveyed to a storage area.

Materials handling also requires the most paper shuffling of the sectors. Materials inventories must be continually accounted for, including workpieces that are between processing sectors. This is a very labor-intensive job for the white-collar workers who keep track of the items that can be in a factory at any given time, as well as keeping sales and marketing abreast of the current levels of inventory available. Figure 2.2 illustrates the interrelationships among sectors throughout a factory, from initial design to product shipment.

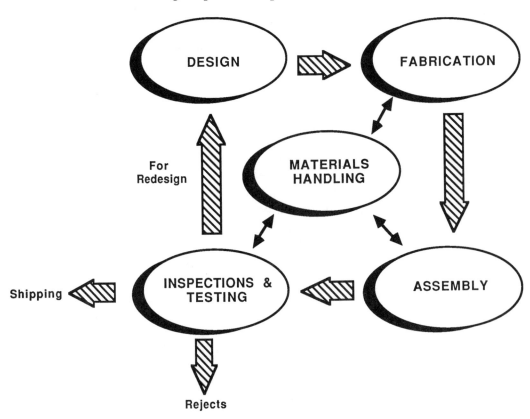

Figure 2.2 Interrelationships among sectors in a factory

Inspection

This sector of manufacturing is perhaps the most important of all of the islands. Here, products are inspected and tested to make sure that they meet the original specifications and work in the manner dictated by the designers. This sector is often referred to as quality assurance checking. Inspection of a product or component usually occurs after the fabrication and assembly steps are performed, so that any defective or improperly fabricated or assembled parts or components can be rejected. The system is designed to work smoothly; however, when the sectors of fabrication and assembly produce backlogs, inspection procedures often skip steps or are performed sporadically in an attempt to eliminate a bottleneck in the production process.

This sector combines the skills of humans and machines in a manner that tries to test a component or product for worthiness. For example, Chrysler Corporation introduced rigid vehicle inspection and testing procedures in 1980 (see "Testing and Inspection at Chrysler") to make sure that their new K Cars were the best that they could be. There are many electronic machines and devices that are used in inspection and testing, ranging from oscilloscopes for inspecting electronic products like TVs and VCRs to specialized machines and equipment that is designed to check a component's or product's tolerances, accuracy, and precision.

The most widely used inspection device is the human eye. There is no substitute for visual inspection. Humans have the capability to spot apparent defects in products and can totally eliminate the necessity for further inspection and testing with machines or other electronic devices. Sadly, human visual inspection is dependent upon the attentiveness of the human. The person who sits all day visually inspecting cabinets used for televisions may let a few obvious product flaws slip by because of a headache, eyestrain, or boredom with the procedure. This is why many products in the American marketplace have a tough time competing against foreign products with higher inspection and testing standards.

In factories that mass-produce products or parts, inspection is often limited to selected batches or runs, because it is impossible to test and inspect every single item that is being produced. A factory producing hundreds of thousands of fan blades will test and inspect only a representative number of each batch or production run of the product. This, hopefully, provides the manufacturer with a picture of the quality of the entire batch or run. It can also mean missing a lot of defective products.

Testing and Inspection at Chrysler

Ever since the early days of Chrysler, engineering and product quality had been perceived by the American public as the area where Chrysler-produced automobiles and trucks were better than the competition, but by the 1970s the reputation began to tarnish. In 1980, the beleaguered Chrysler Corporation, under the helm of Lee Iacocca, instituted tougher inspection and testing procedures in order to reverse the trend of owner dissatisfaction with vehicles that were produced in the late 1970s. These poorly designed, assembled, and tested vehicles were threatening the reputation of quality engineering that Chrysler had held over their competitors for decades. To counter these devastating effects, Iacocca instituted groups of employees called quality circles in Chrysler manufacturing plants. These circles of employees had the responsibility for designing the inspection and testing procedures for the new line of automobiles and trucks that Chrysler was producing. The goal was to make sure that every vehicle that came off of the assembly line was truly road ready. This was to be an especially challenging task because the vehicles that were being produced were entirely new to the Chrysler line. They had no previous record of performance from which to draw problem or performance inferences. The circles debated within each group and finally established the testing and inspection programs. These were called "car evaluation and reliability tests" and would be the final quality checks made on every vehicle before the factory approved it for sale to the American public.

A new testing center was built specifically for the performance tests that were made upon the Chrysler Imperial, a new luxury car for the 1981 market. The quality tests for the Imperial were made adjacent to the Chrysler assembly plant in Windsor, Ontario. After each Imperial had been assembled, it was brought to the testing center. Each car underwent countless inspections and tests, all with a warm engine to replicate actual driving conditions. Vehicles received total underbody fluid-leak inspections and front-end alignment tests and were then given driving tests over an incredibly vigorous, road course that included types of terrain and driving conditions that the car was sure to face while in use. Any vehicle that did not meet the criteria was returned to the factory for corrective measures. Not a single Imperial was approved for sale to the public until it had passed every inspection test and was signed off by the inspector or road tester. Chrysler was able to initiate and perform this elaborate testing upon the Imperial, because it was going to be produced in a limited quantity of only 25,000 units.

The inspection and testing that was necessary for the mainstay of the Chrysler line, the Plymouth and Dodge K Cars, was as rigid, yet much different, because of the number of these cars that were to be manufactured, estimated to be over 600,000 vehicles. The K Cars came in different body styles, and each type was thoroughly tested. Testing included a 5,000-mile driving test (nonstop) that was representative of over 12,000 miles of actual use. The test on

(continued)

the cars was conducted at the Chrysler Proving Grounds at Chelsea, Michigan and consisted of two-lane blacktop roads that included just about every type of road condition imaginable. The cars were also tested for performance on gravel roads, hills of different grades, curves, and loose stones. They underwent extensive testing that simulated high-speed highway driving on concrete test tracks. After testing, engineers checked each vehicle for rattles, as well as inspecting brakes, transmissions, and other areas to determine the effects of the severe testing. Results were sent back to design and engineering so that any modifications required could be made before production actually began. Such extensive testing had not been so thoroughly conducted by any U.S. automaker prior to this time. Overall, the actual final testing of K Cars was performed on selected vehicles from batch assembly lots. The tests were very similar to those that were made on the Imperial and represented an entire assembly run. Chrysler was very successful with this type of batch testing, because they had implemented so many new inspection and testing procedures throughout the fabrication and assembly stages of the production of these cars.

The Manufacturing Environment

The relative ineffectiveness of our current manufacturing methods is being blamed directly for the increasing number of imports, the trade deficits, the high costs of goods, and, of course, the record high unemployment in the U.S. today. There is unrest and loss of prestige in the factory environment, replacing the pride in manufacturing that was so evident 20 years ago. Our current products are becoming notorious for having poor workmanship and unpredictable safety records. Our productivity is at an all-time low, and labor itself is stagnant. Employment possibilities in the manufacturing environment are very bleak; not only are there fewer jobs now available, but those that do exist do not draw the interest that they once did. The young people of today are familiar with the horror stories of corporate manufacturing and are staying away from the manufacturing arena like never before. Unfortunately, unless the manufacturing environment greatly changes, the outlook for future generations wanting to enter this segment of the American workplace does not appear bright. Who wants to work in an environment where there seems to be a subhuman quality to the working life when other exciting opportunities abound? The factory environment in the U.S. has to change, or it will die a slow, painful death. Our approach to manufacturing needs new revitalization, both in procedures and in managerial thinking.

The manufacturing world has become very complicated in the past decade, with many influences affecting it directly. The current external forces that affect the manufacturing world seem to make the operation of a successful industry an almost impossible task. These external forces are caused by the new environment that surrounds manufacturing. They represent changes in our society as a whole, from technological changes and foreign influences to governmental regulations. These forces have a direct impact on the once stable world of manufacturing. As if the external forces were not enough to disrupt manufacturing, internal pressures fuel the already burning fires. Unsatisfied consumers, irate stockholders, confused management, and a largely unhappy work force provide a negative impact on the manufacturing operation. To top off the picture, there is more competition, both domestic and from abroad, than ever before in the manufacturing world. Competition brings with it a wider range of products and prices, as well as the challenges of integrating the manufacturing process among product lines. This new competition has produced substantial changes in the ways companies advertise. There is more advertising today, and more advertising eats into already narrowing profit margins. It almost seems an impossible task to turn around.

Years ago, the traits of a successful manufacturing operation were to:

- Produce goods at a lower cost than competitors.
- Provide consistent quality in all products.
- Meet delivery obligations.
- Continue to provide for research and development, so that new products could be produced quickly and efficiently.
- Satisfy corporate management and stockholders.
- Maintain equipment and facilities.
- Keep inventories at modest levels that were cost-effective.
- Provide employment and job security for the work force.
- Be able to adapt to changes in the environment, both internal and external.

Today, these traits are still valid, but they are becoming more difficult to attain. These requirements often appear to be in conflict with one another in today's industry. No matter what industry does, there are requirements that cannot be met. These requirements sap the resources of the most innovative and ingenious managers; no matter what they do, there is always someone who is not happy with their decisions. It is a no-win situation, to produce at acceptable cost and high quality while being conscious of the company's investment. This is inherent in the operation of manufacturing.

In addition, external pressures add more requirements, so that even the best managers feel as if they did not have the proper tools to cope with the problems. Manufacturing managers often feel cut off with no place to turn for help. This is a result of the nature of manufacturing itself and because of its level of importance in the factories of the 1950s, 1960s and 1970s. During these decades, manufacturing corporations concentrated on continued growth and expansion of market share. Because of this, corporate boards usually paid little, if any, attention to the manufacturing sector of a company (unless problems arose, and this was not the kind of attention that was needed), while they channeled just about every available resource into sales and marketing. Top management was almost totally dominated by decision makers who had expertise in marketing, finance, and sales. The manufacturing managers were much like the cog in the wheel; they simply carried out the decisions from directions provided by the other areas of the corporation, and, like the cog, they may have squeaked, but a little oil soon quieted them again.

This was especially amazing when you stop to consider that the manufacturing sector of most corporations is responsible for about 75 percent of the corporation's total investment, 80 percent of the personnel, and more than 85 percent of the firm's expenditures for materials and other equipment. To imagine that top management paid so little attention to this sector seems almost unthinkable. It's like taking the time to clear a patch of land to plant a garden and, after the planting, not taking the time to water it.

This is what has happened to manufacturing. The irony was that the top management, as well as many industry experts, had mistakenly regarded manufacturing as a stable, almost routine operation that could function with little interference. As long as competent managers were hired, the operation would and should run without mishap. Through the 1960s, as this concept spread, manufacturing was perceived as a dull place of employment and, as a result, did not attract the new generation of managers, workers, and support people. This innovative blood went into other, so-called exciting, arenas like marketing and sales. In the end, the manufacturing sector was left with old management, old ideas, and little future. The downturn had begun. Manufacturing was turning into the black sheep of the corporation. It was necessary, but it received only as much attention as was required. Instead of taking the proper approach of using manufacturing as a competitive tool that allowed a company to produce goods in a short amount of time with low cost and high quality, and where new products could be introduced quickly and be produced in volume with lower costs than the competition, manufacturing stagnated.

Technology, competition, and society itself all bring problems to the manufacturing environment that are not about to disappear anytime soon, but will only increase in the coming decade and force even greater changes in the manufacturing environment. The old concepts and methods for manufacturing are out-of-date and will not work in the information age. There must be changes in the manufacturing community as a whole, changes in skills, equipment, and thinking.

The New Manufacturing Environment

To regain manufacturing competency, industry and its management must take the initiative to break the bonds of yesterday's methodology. A new manufacturing environment needs to take shape in order for the industry to recover and thrive. Instead of the traditional short-term approach of solving manufacturing problems by considering only production and profits, corporations must now start to think of the long-term potential and consequences of manufacturing operations. What works today might not work tomorrow. Manufacturing must be more conscious of quality, service, and have the ability to change to meet the requirements of internal and external forces while continuing to focus on productivity and efficiency.

To succeed, manufacturing must share its focus on productivity and efficiency with important issues that directly affect its everyday workings. These issues include the number and size of its factories, the choices in equipment and technology that produce the most "bang for the buck," and the mechanism of controlling the entire operation in a way that maintains the continuity of manufacturing and corporate success. Manufacturing is only a component of a corporation, and it must share in the long-term goals and desires of that corporation. There has to be a direct link, so that both entities work together for the common good. In this area, control is the key, and information exchange provides the necessary control. Improving the control process furthers the ability of manufacturing managers to anticipate and make strategic decisions that link all sectors of the factory and corporation together.

Corporate strategy and manufacturing strategy must share resources and provide for a common, attainable goal. Sound strategy will eliminate the making of piecemeal decisions on the fly. The establishing of a sound manufacturing strategy, in most cases, eliminates many trivial decisions, because it provides more focus and understanding of the tasks that must be accomplished for the corporate good. This process alone benefits industry, because it prevents the loss of time and money while waiting for managerial decisions to be made. IBM, American Airlines, and General Motors have established sound long-term strategies, and their managers work daily to see that the strategies are followed.

They have adapted strategies based upon technological change, and their corporate strategies reflect this fact.

The coming changes in technology will affect society as a whole and will certainly affect manufacturing as well. As society accepts these changes, so too will the manufacturing environment. The changes will cause new production methodologies to surface and new attitudes of management to be born. This will be true of other sectors of business as well. Just as the world of the office was changed by technological advancements, so too will the factory change.

Instead of being managed from the bottom up, manufacturing will reverse the process and manage from the top down, like other businesses. The bottom-to-top philosophy has prevailed for so many years that, to many, changing it is almost unthinkable. Frederick W. Taylor, long considered a guru of manufacturing principles, wrote about the bottom-to-top process many years ago. In this time-honored process, which served the age of mass production, industry would select a manufacturing operation and break it into its unique elements. Each element would be thoroughly studied, and methods for improvement would be devised and implemented. Then, the elements would be reassembled into the original operation. This approach was documented in numerous texts and studied by managers and those desiring to be managers. As a result, this philosophy prevails in industry today.

To replace this bottom-to-top process and to provide for more efficiency, manufacturing should adopt the management style of top-to-bottom techniques. This process is much more productive and is in step with today's markets, as it fosters shorter production runs and greater flexibility, adapts more easily to technological or product changes, and is designed for the competitive nature of today's manufacturing. By implementing the top-to-bottom approach, manufacturers can respond much faster to social or consumer demands. The approach is a derivative of the corporation's competitive manufacturing strategy, which defines a company's manufacturing policy. The strategic policies provide the necessary instruction so that the factory's technical experts, managers, engineers, and computer specialists can accomplish their jobs more efficiently. It also provides a method for upper management to monitor and evaluate a company's overall performance by meeting objectives within the policy statement. This approach is more effective than merely monitoring overall production statistics. Because it focuses upon the corporation's strategy and the policies of implementing the strategy, the manufacturer obtains a higher degree of control over the overall operation of the manufacturing environment and becomes more competitive. To make the most of the top-down approach, however, manufacturing still needs one more thing, automation.

Managers in the Factory

For centuries, the factory environment has been subject to clashes between labor and management. From the time of the earliest factories in Great Britain to the labor struggles that occurred in the United States during the 1930s and 1940s, the workers of industry have banded together to question the decisions and abilities of management. Labor continues to cast a doubtful eye toward manufacturing management even today, and undoubtedly will look hard at the issues involving computer automation and robotics technology.

Today's factory workers, for the most part, are an unhappy group; many feel that they are stuck in a situation where they are valued only for their labor skills. Granted, factory workers do earn a substantial wage (for union workers especially); however, the tasks which they perform are generally jobs with little hope for advancement or career fulfillment. They stick it out, because of economic necessity and the lack of other jobs with comparable earning levels. Besides, factory workers have little in the way of marketable skills that are useful outside of the manufacturing environment. As expected, the morale in the factory is at an all-time low, the workers themselves exhibit a complacent attitude toward their work that shows in the quality and quantity of goods produced. This makes it difficult for the most competent manager, while it creates an almost impossible task for the mediocre manager. Automation will not necessarily eliminate these problems.

The coming automation of our factories will create more changes in the methods of manufacturing and management than have occurred during the last century. Computer automation and robotics will make our factories more flexible and productive, but the implementation of these technologies will require major investments of time, money, and, above all, total commitment of top and middle management. The commitment of management is paramount, as it will be management that provides the direction, confidence, and leadership necessary to make automation succeed. Progressive companies that are now realizing the benefits of automation and robotics have already taken the necessary risks and made the required investment in computer-automated equipment. For them, the commitment is already established. They now must incorporate new strategic policies to make sure that they make the most of their commitment and their investment. It is well worth the effort. To achieve the dramatic increases in productivity and flexibility, there is no other alternative available today that has such potential to completely turn around the stagnating manufacturing industry in America. But, the majority of manufacturers are not using, nor even considering the use of robots and computers. Fewer than 7 percent of America's manufacturers use computer automation.

Corporate management is unwilling to anticipate and accept the changeover to automated techniques. It is the nature of manufacturing management to look to short-range solutions to problems rather than long-term solutions. The short-range mode of thinking centers around the productivity and cost issues that have been expounded as manufacturing management doctrine for the past 50 years. It is a tremendous obstacle to overcome, but it is not insurmountable.

For all of its possibilities for manufacturing improvement, computer automation is not a task that is easily implemented; managers realize this, and it scares them away from the technology. It is a sound fear; spending hundreds of millions of dollars to completely retrofit a factory is an expensive proposition. It is well worth the risks and investment involved, because it is the only method that will rejuvenate our manufacturing processes and make us competitive again.

The decision to automate requires total commitment and an entirely new way of thinking before the conversion from present manufacturing techniques to computer-automated techniques can be successful. This is very similar to the situation in the office environment when computer automation first started to appear there. Many business managers resisted office automation, thinking it could not help them or that they could not afford the new technology. They remained paperbound for several years until they realized that using a computer to automate the office would ultimately make them more productive and provide them with a competitive edge over other businesses. With this enlightenment, office automation grew.

Today, the use of office automation equipment is growing steadily, and the productivity of the paperless office is increasing, while the number of managers resisting automating the office environment is declining. The productivity and competitive gains are too impressive to be ignored, and management realizes this. The same process is starting to occur in the factory. Manufacturing managers are becoming conscious of the competitive edge manufacturing automation can provide them. But more of these progressive managers are needed, and they are needed soon.

The Automation Manager

The conversion from traditional manufacturing methods to the computer-automated/robotic operations of tomorrow will not be limited by economics or by the lack of computer technology. It will be hampered and restricted only by the skills of management and the management systems that are currently in place in our factories.

Computer automation and robotic implementation will require an entirely new breed of managers, because technology will change the role of management in the factory. Automation will transfer managerial power from the traditional process-oriented manager to the new manager who is experienced in planning, training, and communicating, and familiar with technology. This new manager will be anxious for change, replacing the old values of current management who have the "let the other guys take the risk, I'm certainly not going to" attitude. The old guard upholds the logic that investing in the new automation is foolish when the current methods, though antiquated, still keep the factory running. Resistance to change is a common trait of the current managers of our factories; they are traditionally very conservative and avoid taking risks at almost every extreme.

The new managers of automation will possess skills and cognitive thinking characteristics that will enable them to conceptualize ideas and put the ideas into motion much more efficiently. They won't rely upon past performance to discern future needs. They will focus upon corporate and manufacturing strategy rather than production/cost-efficiency. They will adapt easily to the top-to-bottom style of managing that is necessary for automation to succeed.

These new managers will understand the need for flexibility on the factory floor and to adapt to the different forces of their environment quickly and readily. The new managers will possess knowledge of electronics, mechanical aptitude, and the science of management, and have a keen awareness of the overall business environment, especially the areas of engineering, marketing, sales, personnel, and administration. These new managers will understand the importance of communication among sectors of the factory. They will encourage the sharing and promotion of new ideas and concepts. They will have innate ability to communicate and exercise control without being feared or loathed. They will understand how to work in a group environment and be particularly skilled in labor relations—this is a must. The new breed will have the ability to perform sound planning tasks and maintain superior project control while managing.

Admittedly, only Superman could possess all of these traits;

yet, the new managers will likely possess many of these intrinsic qualities. They will bring new experiences and ideas to the factory environment, because they have come up in the ranks from the other corporate areas of research and development, marketing, and sales, instead of climbing the more traditional career ladder of manufacturing—from the assembly line to the board room. Many of these people will have a thorough understanding of computer technology, and many will have already used computers to accomplish their jobs in the other corporate areas. Some individuals will even understand computer programming and have developed programs of their own. Many will have extensive data-processing management backgrounds, where they have implemented computer systems and coordinated operations for the users of these systems. A key principle will be their capability to analyze problems quickly and effect solutions in a logical manner, and with as much automated support as is possible.

As computer automation increases in industry, there will be many new managers who will join the manufacturing field, coming directly from managerial positions in industries other than manufacturing. These workers will come from companies that have exceptional track records in many different areas, providing the manufacturing community with new ideas and methods of successful business operation. These people will be an asset to the manufacturing community, because of their fresh ideas, attitudes, and knowledge.

Trends in Education

In colleges and universities across the U.S. today, there is increased interest and enrollment in courses in manufacturing and the applications of robotics in the factory. There is a shortage of people with experience and familiarity with robotics, but this is changing. More and more students are attending classes in mechanical, electronic, and industrial engineering. To serve this renewed interest, industry and education have developed programs for continuing education and research in the areas of computer automation and robotics. For example, Cincinnati Milacron, a domestic robot manufacturer, has contributed several TR3 robots to Ohio State University and Georgia Tech University for educational purposes. IBM has done the same thing, providing over $40 million in grants for computer-aided design/-computer-aided manufacturing (CAD/CAM) research at over 20 schools. In addition, robotics and computer-automation studies are taught at many leading schools across America; these school include Carnegie-Mellon University, Stanford University, M.I.T., Purdue University, University of Texas, Austin and Arlington, Lehigh University, Macomb County Community College, and the

University of Rhode Island. In addition, many of these institutions of higher learning also offer coursework in business applications that directly relate to manufacturing and industrial automation. All of these classes will provide the manufacturing community with more graduates who possess an understanding of computer automation and robotics and their relationships with industry than have ever been available before.

What will be the overall effect? Automation will make the factory of tomorrow a more exciting challenging place. The old mundane tasks will be replaced with automation, freeing managers to manage, while making the tasks of the work force more challenging and satisfying. It may take a few years, but the generation of young people that is currently studying these interesting disciplines will make the concept of the factory of tomorrow a reality.

CHAPTER
3

Computer Automation, Robots, and the Japanese

The world of robotics and computer automation is, for the most part, new to many individuals. And, by being a part of the computer industry, the two technologies have a vocabularly all their own. They also have specific applications that are unfamiliar to most people, even to some in the manufacturing community. This chapter introduces some basic vocabulary and discusses computer automation in general. It also provides information describing computer-automated robotic applications in the United States. Finally, it discusses the Japanese, and their use of computers and robots—they are the real innovators in the use of this manufacturing technology.

Computers

A computer is a device that electronically stores information or data. This information comes in a variety of forms: it might be numbers, words, or even graphic images. Information is put into a computer in a number of ways, by using a typewriter-like keyboard, a light pen pointer, an optical scanner capable of reading written words, a touch screen, and even voice commands. Once data is entered into a computer, it can be manipulated and changed according to a set of mathematical rules that provide instructions telling the computer what to do. This set of instructions is called a program.

A program can contain one task, or several different programs can be integrated with one another, with each program performing a different function. Different types of programs perform all kinds of mathematical processes, generate text, create and alter images, or send electronic commands to machines. Computers can do many different tasks that were previously performed by human hands and minds.

Computers first began to appear over 40 years ago. They were very large, often occupying enormous air-conditioned rooms where they consumed vast amounts of electricity and human time, while they methodically processed incredible amounts of information faster than any human could. Needless to say, the cost of these early computers was nearly as large as the machines themselves. Early computers were simply not cost-effective for everyone to use. Only large businesses with extensive capital could afford them. As a result, they were most often found in insurance companies, big banks, and large governmental agencies.

In the mid 1970s, the microprocessor was developed. It enabled computers to process more information by using a tiny silicon chip. Almost overnight, computers became smaller, less expensive, and more powerful. They came in all shapes and sizes and performed many tasks that helped us in our businesses, schools, and homes. Thousands of different programs were developed for just about every application imaginable. Included were applications that could help revitalize the American manufacturing industry—the controlling of machines by computers.

Already, computers had started to revolutionize offices. The microcomputer was being used in businesses across the country with astounding success. Its impact was remarkable. For example, at the New York Stock Exchange (NYSE), productivity increased over 400 percent after the installation of an integrated computer network. The NYSE computers store, track, and retrieve data, representing the millions of daily financial transactions, quickly and efficiently. Before computer automation, the NYSE's paper records usually lagged hours behind the transactions that were made during a normal business day. Active trading days were a paper shuffler's nightmare; final results were sometimes delayed until the following morning, while information was manually manipulated and posted. Today, more than 10 million computer terminals are used to conduct America's daily business. From banks to hospitals to supermarkets, business people use computers to perform just about every type of information-processing activity that was done by hand as late as 1975. Incredible as it may sound, two out of every three office workers use some type of information-processing equipment in their daily work. As a result, business is more productive.

The computer replaces the old manual methods of exchanging information (paper shuffling), because it can send and receive information instantly. Instead of taking hours to record and manipulate information, computers process and exchange information in seconds. From the largest corporation to the smallest single proprietorship, computers increase productivity and help managers achieve more control.

The computer industry quickly became an important part of our economy, because of its far-reaching impact. However, there were still areas that the computer hadn't touched. During the second half of the 1970s, many changes affected our country. We began to to see a flood of foreign goods coming into the United

States to satisfy our consumer market. A large number of these imports were electronic goods, automobiles, and clothing. American industry perceived this increasing number of foreign goods as a threat to its economic position. Unfortunately, while American consumers almost seemed to gobble up these less expensive, high-quality imports, relatively ignoring competing domestic wares, our trading partners' wallets grew fat.

While we bought foreign goods, we watched American industry suffer, without really caring. Americans were tired of paying high prices for products that were known for inconsistent quality and questionable workmanship. Not only did manufacturers suffer the losses due to goods not sold, they found their coffers drained even more by the incredibly high wages of the work force. It certainly didn't help their situation when they had to spend millions of dollars on product repairs that were required for warranty work on defective products. Even advertising didn't help. Americans were simply not as interested in domestic products as they had been in the past.

As the economy worsened, manufacturing executives tried to figure out what had happened. The answer was easy; our factories chose not to change with the times for almost two decades. While technology advanced, manufacturing slept. Unfortunately, the short nap almost became a complete hibernation.

The Robots are Coming

By the late 1970s, thousands of businesses were either sick or dying, most being victims of the dreaded import. The malaise affected all of America, causing extreme economic downturns in many sectors of American commerce, particularly those in manufacturing. Many businesses and factories quietly closed their doors voluntarily; they simply gave up and called it quits thinking that they could not compete against the flood of foreign goods. Others, including many manufacturers of clothing, automobiles, electronics, and lumber, were very close to being forced into bankruptcy by their creditors. However, these manufacturers had the desire and the determination to fight.

What had occurred during the past two decades that made our manufacturing methods so obsolete now? It was mostly our complacent attitudes. While we didn't significantly adapt new technologies into our factories, our foreign competitors did. While we didn't bother to improve our out-of-date management philosophies, our foreign competitors did. While we didn't think it wise to invest in new, more efficient machinery and equipment, our foreign competitors did. In order to succeed, American industry has to play the game better than they have in the past. U.S. industry must start using the same resources and tools that their competition is using. The tools are the new computers, and the resources are the new philosophies and strategies for improving our manufacturing methods.

In the late 1970s, leading business experts began trying to identify and solve the many complex problems and situations that were having such dire effects upon the American factory. While they analyzed the manufacturing environment, breaking it down into its components, the experts realized that lack of control was a major area of concern in the factory environment. This lack of control was evident throughout the majority of the companies that were experiencing major problems. As strange as it may seem, American industry was operating in the late 1970s with much less control and integration than was present in the other areas of the business community. After reviewing the management techniques being used in the decision-making management processes of these companies (management techniques that dated back to the 1940s), it was apparent that the old concepts just were not designed to be used in the modern factory. Technology had made them obsolete, just like the manufacturing equipment that they were still using. The business experts reasoned, that computer technology might be used to help industry become more efficient and productive in the same manner that it had boosted productivity in the office environment:

For the first time in our manufacturing history, a number of leading American manufacturing corporations began to develop strategies that centered around installing and using computers and automated robots in an all-out effort to replace the outdated manufacturing processes and strategies that were crippling American industry. A milestone in manufacturing had been reached. As early as 1979, some manufacturers were already using computer-controlled robots in a number of different applications. The majority of these new robots were installed in showcase factories of large corporations, such as General Motors, Chrysler, General Electric, John Deere, and IBM.

These companies were the industry leaders that took the necessary risks and made the enormous expenditures in order to acquire this new automation that could possibly make them more competitive. They committed themselves early to the new technologies and to the benefits that might be attained by using computers and robots in the manufacturing sector. The companies that were the first to use the new tools shared common traits. They all had much to lose from foreign competition, they all had the money to invest in automation, and they all had the desire to improve. It was a bold move, even for the largest and most financially secure corporations, and it required progressive and insightful managers who were willing and able to make hard decisions that would affect their companies for years to come. It was not an easy task. After all, investing hundreds of millions of dollars incorrectly could very well irritate many stockholders. It could also bankrupt a corporation. However, realizing that the opportunities far outweighed the possible risks, these modern-day pioneers began to install computer-controlled robots in their manufacturing plants.

Robots are not of themselves the actual key to factory automation. They are a peripheral device, controlled by the computer itself. The programs that are stored in the computer's memory instruct the operations of the robot. The computer can control an indiviual robot or a group of robots, depending on the sophistication of the program. The computer could even be programmed to send instructions to the processing machines that are used in the manufacturing cycle. Slowly, computers began to enter the world of the factory. As technology and time progressed, both the robot and the computer evolved, and where we once considered robots only the dreams of science fiction stories, we now think of these technological wonders as a normal part of the manufacturing world and, certainly, a bigger part of our future.

What are Robots?

Robot is a Slavic word. It was first used in our language in 1922, coming from a play called *Rossum's Universal Robots*. Robot means heavy work or forced laborer. In common usage today, a robot is a mechanical device that has a fairly high degree of intelligence and the capacity to perform different operations when connected to a computer. It is programmable in the sense that the computer program controlling the robot can be modified to accomplish diverse tasks that were originally conceived in human terms. The discipline of conceptualizing, designing, building, and applying robots is called *robotics*. This term was coined by science fiction author Isaac Asimov in his 1942 story, *Runaround*.

A robot is a computer-programmable, mechanical device capable of performing work. There are no robots today that are as brilliant and agile as the gold-plated electronic wonder that accompanied Luke on his adventures throughout the galaxy. Just the opposite is true. The robots currently in use are very plain-looking machines. Most look very awkward, yet they can manipulate their humanlike arms and fingers in a variety of motions. These fluid moving robots are true symbols of the extent of the processing power and control provided by the computer.

In general, robots have parts that are analogous to the human body: hands, arms, joints, and a brain. A robot's hand is called an *end effector*; its arm, called a *manipulator*, is made of a shaft with a flexible joint at the wrist, elbow and shoulder. The shoulder is where the body of the robot usually ends, and it is used to attach the arm to a sturdier fixture. The shoulder is usually mounted on a pedestal or hung from a gantry. Some robots are mounted on pedestals with wheels or rollers to allow independent movement of the robot. The robot's brain is the computer; it is as powerful as the limits of the computer itself. The previous description applies only to one type of robot, one that is reasonably sophisticated, both in design and capabilities. Many robots are much simpler and look more like machines, and they do not have parts that look remotely humanlike. Several types of robots are illustrated in Figure 3.1. A variety of robots and their uses are more fully described beginning in Chapter 5.

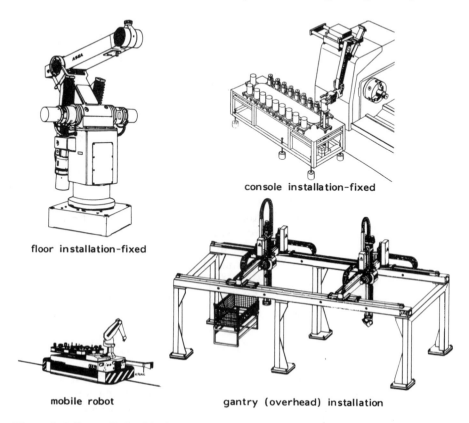

floor installation-fixed

console installation-fixed

mobile robot

gantry (overhead) installation

Figure 3.1 Types of industrial robots

Robotic Terminology

Since robotics has a unique vocabulary, the following paragraphs define a few of the common terms.

Degree of Freedom

One of a limited number of ways a robot can move. For example, a robot arm mounted on a base can rotate, often up to 270 degrees, it can pivot up and down, and it can telescope. Rotation, tilting, and extension/retraction are its three degrees of freedom and delimit the robot's capabilities. The wrist is capable of three further degrees of freedom. It can bend in a vertical plane, swivel horizontally, or rotate on the arm. This gives six degrees of freedom, but fewer are quite often adequate. Naturally, the simpler the robot, the fewer degrees of freedom; the more complex the robot, the more degrees of freedom. Two examples of degrees of freedom in industrial robots are illustrated in Figures 3.2 and 3.3.

End Effector

The end effector is, in plain English, the robot's hand. Like a human hand, the end effector is used to grasp objects, perhaps a workpiece or some other part. The end effector must be able to grasp and lift a workpiece without damaging or dropping it. Robot hands are extremely simple compared to human hands. In many cases, the end effector is not a hand at all, but a detachable tool, such as a paint spraying gun, a grinder, or a welding gun. Types of gripping actions that are designed into robot end effectors are shown in Figure 3.4.; Figure 3.5 shows a typical end effector.

Manipulator

The term manipulator refers to the robot as a whole, including base, power supply, and arm. More specifically, it is the mechanical arm that actually moves itself and its workpiece through space. A typical manipulator is shown in Figure 3.6. Its motions are described in relation to a coordinate system, either rectangular, cylindrical, or spherical.

Controller

The controller is the brain and nervous system for the robot. Any programmable device can serve as a robotic controller, whether it consists of mechanical stops, a rotary drum switch, or the most advanced mainframe computer. The controller is basically an information-processing device with inputs of desired and measured position, velocity, or other variables, and

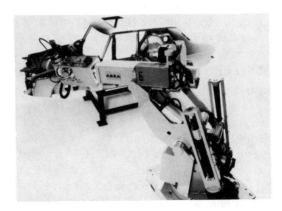

Figure 3.2 A spot-welding robot with six degrees of freedom, and a symbolic respresentation.

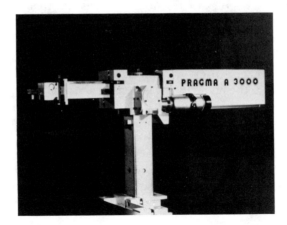

Figure 3.3 An assembly robot with three to six degrees of freedom, and a symbolic representation.

whose outputs are drive signals to a controlling motor or actuator. The most sophisticated controllers direct the robots through their paces, integrating them with other equipment and keeping watch on the overall system. They do all this while making system-relevant decisions and reporting the system status to operations control.

Coordinate Systems

The method for determining the work envelope or spatial volume in which the robot operates. The main configurations in use today are the following:

1. Cartesian, or rectangular, coordinates, defined by three distances, in which the work space resembles a rectangle;
2. cylindrical coordinates, defined by two distances and an angle, in which the work space resembles a cylinder;

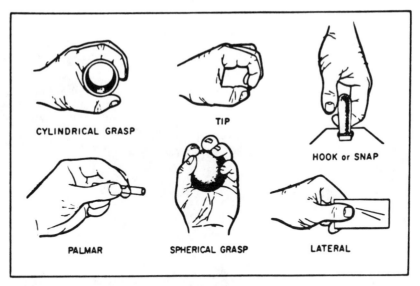

Figure 3.4 The various types of hand prehension desired in end effectors

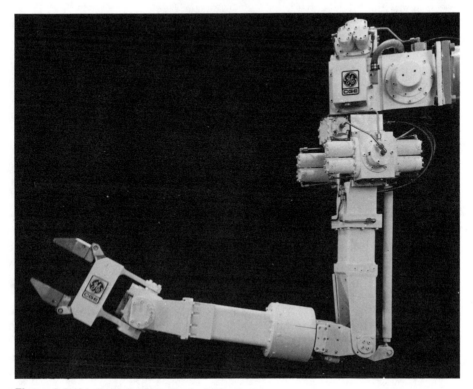

Figure 3.5 A typical robot arm with attached end effector

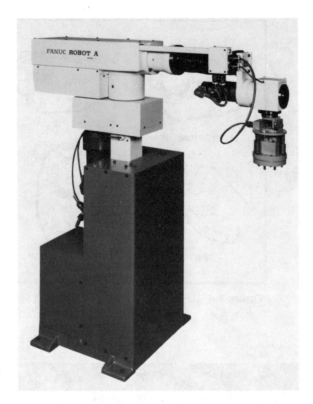

Figure 3.6
A typical manipulator, including the base, power supply, and arm

3. spherical coordinates, defined by two angles and a distance, in which the work space resembles a sphere;
4. jointed arm configuration, which combines the features of the other systems. Figures 3.7 through 3.10 illustrate common work envelopes; these are determined by the specifications of the machine's coordinate programming. In each figure, the shaded area illustrates the work envelope defined by the coordinates.

Robotic Efficiency

Some robots are capable of doing many jobs, but most are capable of several jobs within a range and with imposed limits. The limits are set either by the computer providing the control or by the physical limitations of the robot itself. For example, one of the most popular applications for robots is spray painting. In this job, a robot that is designed to paint cars can be programmed to paint many different types of cars, but the robot requires substantial changes to be able to do anything but paint.

The real difference among robots is determined by the number of different tasks or operations that the robot can do. The industry term for a robot's adaptability to efficiently perform a

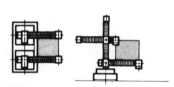

Figure 3.7
Rectangular coordinate robot

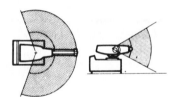

Figure 3.8
Spherical coordinate robot

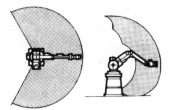

Figure 3.9
Jointed arm robot

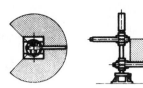

Figure 3.10
Cylindrical coordinate robot

variety of different tasks is flexibility; flexibility is achieved by using sophisticated robots that are capable of performing different operations through the reprogramming of the computer instructions. For example, some industrial robots are flexible enough to perform pick-and-place functions as well as loading /unloading tasks; in order for the robot to perform both tasks, the program that controls the robot has to contain information that describes both operations. If the robot was originally used to load and unload pallets containing workpieces, before it could be used to do a different task, some modifications would have to be made in the original program.

Reprogramming provides the flexibility that allows the machine to accomplish different functions. A variety of different tasks can be accomplished using a single robot, which in turn increases the cost-effectiveness of the machine. The reprogramming capability adds new dimensions to the robot's overall potential to perform work and is similar to the limited retooling effort that is required in today's manufacturing plants. This flexibility can translate into the production of small numbers of specialized items with all of the benefits of economies of scale that are achieved in larger, mass-produced runs. Although the initial cost of such a system is high, the savings afforded by the ability to make quick programming changes are significant; robot operations can be altered as needed, using existing equipment, and the changes in robot functions can be achieved without interrupting production.

Types of Robots   The majority of robots at work today are industrial robots. There are several ways to classify robots: by the controller and drive description, by the kind of work done, and by the degree of complexity. Industrial robots can be further divided into two groups: servocontrolled robots and nonservocontrolled robots. A servocontrol mechanism is an automatic control device that reacts in response to a feedback signal which is a function of the difference between the command position and/or rate and the measured actual position and/or rate. Servorobot actions are controlled by feedback to the controlling device. Nonservo-robots are less expensive and simpler in design and operation, but considerably more limited in scope than servorobots.

Degree of Complexity

Industrial robots are available in three levels of intellectual capacity and performance ability: simple robots, medium technology, and sophisticated industrial robots. Simple robots, also called pick-and-place or limited-sequence manipulators, are low cost, easy to keep on-line, and offer significant increases in productivity by virtue of their speed and accuracy. They are limited in information memory capacity and usually restricted to two or three nonservo degrees of freedom. Simple robots are dependent on ancillary equipment, such as bowl feeders, because of their limited number of available moves. This is the type of robot that the Japanese use most in their factories.

Medium technology robots have larger memory capacities and are more easily programmed to perform a wider variety of tasks. They have up to six degrees of freedom and are usually servocontrolled. They have multiple-step memory and programmability in their major axes, i.e., wrist rotation, and radial and vertical transverse. Usually used in single-machine load/unload situations, they are incapable of continuous path functions, such as welding or spray painting. Many areas in modern manufacturing remain open to the use of these robots.

Sophisticated industrial robots are considered to be the leading edge of design/manufacturing technology. These are what could eventually evolve into the classically romanticized androids of near- or superhuman abilities. They are physical extensions of the highly sophisticated computers that control them. They have flexible, programmable manipulators and embody the highest degree of artificial intelligence in industrial automation. They are readily programmable and can easily be integrated into a wide variety of workcells and systems. Sophisticated industrial robots have considerable memory capacity, which enables them to store and change many programs.

Control/Drive Description

Another scheme for classifying industrial robots is to describe their drive systems, or the method of actuation. Those in use in virtually all modern industrial robots are pneumatic, electromechanical, and hydraulic drives. Pneumatic drive systems are found in about 30 percent of industrial robots. They use standard vane motors and cylinders to extend, retract, or rotate a joint or arm. These drives are relatively inexpensive and quite reliable. They have moderate load capability and torque capacity. On the negative side, it is not easy to control the speed or position of the robot, which is an important requisite.

Electromechanical drives are in use in over 20 percent of industrial robots. Servomotors, or stepping motors, pulse motors, and linear or rational solenoids, are used to obtain a high degree of accuracy and repeatability. On the other hand, electrodrives have a lower load capability and torque capacity than either the pneumatic or hydraulic drive systems.

Hydraulic drives are used in 45 percent of industrial robots and are the real workhorses of the field. By means of hydraulic motors and cylinders, the hydraulic actuator converts energy from pressurized fluid into shaft rotation or linear motion. This provides a high level of power and force, along with compactness and accuracy of control. Fluid power is simpler, more durable, and more resistant to harsh environments than other drives. While they are currently the most expensive drives, they also provide the highest load capability and torque capacity.

Robot Applications

In the last decade, robots have shown rapid evolution from the drawing board and laboratory to actual implementation. Modern industrial robots are extremely versatile. They have been successfully introduced in such diverse fields as welding, polishing and grinding, gluing, spray painting, and assembling. Robots are best applied in the many dirty, unpleasant, demeaning, and downright dangerous jobs that human beings don't want.

Robots were initially designed for, and got their first jobs, doing work that was either too dangerous or too repetitive to be done by a human being. The first installed robot was used in the hazardous environment of a General Motors die-casting facility, a dangerous place for a human to work. Uses for robots have not changed significantly; they are still used exclusively to perform the most dangerous and distasteful manufacturing occupations. Many robotics experts foresee the day when just about every manufacturing task can be performed more productively and efficiently by a robot than by a person. It is very unlikely that this will really happen; however, robots are expected to replace humans in many manufacturing jobs in the factory of the future. These automated workers hold more promise for the revitalization of our manufacturing industries than any conceivable alternative. As a result, robot use is expected to increase in the coming decades.

Welding, although a skilled job, is a dirty, boring, and dangerous job that requires extensive training. As a result, although it pays well, there is a chronic shortage of qualified welders. Therefore, it is not surprising that welding robots comprise about one-third of the U.S. robot work force today. Spot welding was the first significant application of robots. This is largely due to the increasing reliance of the automobile industry on robotic welding. That industry has a high visibility factor and a declared intent to continue robotization, which has made it a main target for robot manufacturers. Over the last decade, robotic spot welding has evolved from a generalized machine to a specific-purpose unit suitable for large-scale manufacturing. This means that in time their use will become standard for all high-volume spot-welding assembly needs. Arc welding by robots is a consistently growing application. Ultimately, almost all arc welding will be done by robots. However, improved vision systems are needed for seam tracking before maximum potential can be attained. Cost is also a prohibitive factor for some companies, but it is becoming less so with the advent of simpler, job-specific units, such as those built by Yamaha. Other cost factors are increased productivity due to faster welding and more consistent quality control.

Most robotic applications began in large factories, but some significant applications, such as paint spraying, originated in small workshops. Spray painting by robots was begun by Ole Malaug in a small Norwegian company. Malaug initially used his robot paint sprayer to paint small implements, wheelbarrows, and gears, because the painting booth is a production bottleneck. Malaug's company, Trallfa, now dominates the world's spraying robot market with a share estimated at over 70 percent.

Parts assembly seems to offer the biggest challenge and the greatest potential rewards to roboticists today. Assembly accounts for over 50 percent of the total production time for most manufactured goods, but only about 6 percent of assembly tasks are fully automated. Many tasks, such as installation of valves and oil seals, insertion of washers or springs into cylinder heads, etc., are readily performed by robots. Almost all assembly is very simple and can be automated quite easily. So far, the techniques have not been sufficiently mastered to allow widespread introduction. This is largely due to the precise nature of most detailed assembly tasks. Before robots can meet the challenges of full-scale production, more sophisticated visual and tactile sensory systems must be developed.

Nonindustrial Robot Applications

Although at present most robotic applications are factory-production oriented, considerable research and development is being done in other areas. As a result of this research, more robots are being used in areas never before exposed to the technology. As mentioned previously, undersea exploration and agriculture stand to benefit greatly from robots. Space holds many opportunities for robotic applications. The robotic arm of the space shuttle has received considerable attention from the news media for its role in assisting astronauts deploy and retrieve satellites. NASA has even gone so far as to predict that robots will be used to assemble parts of space stations in the near future.

The military is particularly interested in certain areas. Odetics, Inc. of Anaheim, California has contracted with the U.S. Navy to develop an onboard fire-fighting robot. Though the design is still being perfected, the unit is envisioned as moving on wheels rather than on legs. It will be connected by fireproof hose to a large supply of fire-fighting chemicals; it will also carry smaller amounts of other chemicals and be controlled from a remote source. The Pentagon is interested in robot planes, ground assault vehicles, and underwater robots, especially for deep-sea salvage and rescue operations. For all of the dreams and ideas, there are many robots used in nonindustrial applications, and the number is growing. The following paragraphs illustrate some of these nonindustrial uses.

Students at Louisiana State University are working on a simple robot that transfers seedling plants to a transplanting machine. This is a perfect task for a robot, because it is a tedious, repetitive, and well-defined job. It also has tremendous possibilities for practical use in land management, reclamation, and forest management. Robotic harvesters for crops are also under study. LSU is making strides towards the perfection of an orange-picking robot. With this robot, when an orange is identified, the arm can reach straight in and pick it. LSU is using artificial light filtered with a very narrow band-pass filter centered at 675 nanometers to get high-contrast computer images that represent oranges as white blobs surrounded by a dark backdrop of leaves and limbs.

Mining and excavation hold promise as fields for important contributions by robots. The Dravo Corporation of Pittsburgh, Pennsylvania has joined with Carnegie-Mellon University to produce an excavation robot. The excavation robot will boost productivity substantially, permitting much more rapid mapping and exposing of gas pipes than conventional methods. The safety factor is another major consideration in this area since many people are killed in the United States every year in fires and explosions associated with leaky gas lines, mostly during repair.''

A second project involving Carnegie-Mellon is the construction of a guidance system for a coal-mining robot, in conjunction with Denning Mobile Robotics, Inc. of Woburn, Massachusetts. In an old, unsafe mine, the roof needs to be bolted up so it does not fall on the miners' heads. Though the existing system for this task is largely automated, the robot has to be positioned by a human operator. Denning is using modeling and navigational technologies developed earlier to allow the robot to position itself as needed, thus eliminating the need for human involvement in the process.

There are many other applications for robots, actual and contemplated. More are being conceived daily, including dishwashing, sorting fruit or vegetables, candy processing, mail sorting and delivery, meat cutting and grading, reconnaissance, bricklaying, pool cleaning, and asphalt laying.

The Japanese

In 1956, the year of Devol's first automated manipulator, Japan was still recovering from the effects of World War II. The entire country was rebuilding. This provided the opportunity for new industry to be built from the ground up, using new design technologies and the available automated techniques. It also afforded the Japanese the chance to redevelop their manufacturing strategy. Although Japan entered the industrial age later than the United States, this has helped them. By entering the era of manufacturing later, they acquired newer equipment and facilities, especially at the end of the war. Their investment in new equipment and facilities automatically put them ahead of their American counterparts. For example, while the American steel factories continue to use outdated open-hearth furnaces, the Japanese use the latest electric furnace technology and produce high-quality steel at lower costs, enabling them to capture a greater market share.

While we have been slow to implement computer automation and robotics into our manufacturing facilities, the Japanese have wasted little time since the technology became available. As a result, the Japanese have already discovered that using computer-controlled robots in their manufacturing plants is very cost-effective. Today they use thousands of robots in a number of different applications. This is a remarkable achievement in view of the fact that the Japanese first considered the possibilities of robotic automation less than 20 years ago.

In 1967, Joseph Engelberger visited Japan to speak at a symposium on robots and computers. After his formal presentation, which lasted a little over one hour, the several hundred Japanese manufacturing engineers and managers who were in attendance were so interested that the question-and-answer period lasted over five hours. Japan was hooked on the possibilities of this amazing technology. One year later, in 1968, Kawasaki Heavy Industries, Ltd. began using robots in manufacturing as a licensee of Unimation—Engelberger and Devol's company. Kawasaki is now Japan's leader in robotic manufacturing.

By the mid 1970s, Japan had become very agressive in the introduction and use of robots. They discovered that the money they invested in robotics and computer automation would be returned through the increases in productivity and reduced manufacturing costs (excluding capital investment, which is depreciated over many years). In 1971, due to the increasing use and interest in computer automation and robotics technology, the Japan Industrial Robot Association was founded, with an initial membership of fewer than 50 different companies. The Americans did not form such an association (The Robot Institute of America) until 1975. The Japanese now use and manufacture more robots than any other nation. They estimate that productivity gains are 30 percent. The quality of Japanese products has improved drastically from the cheap throwaway products of the 1950s and early 1960s. Robots and computer automation eliminated the problems associated in human error and manufacturing inefficiency, and this led to higher consistency in product reliability and overall quality. The discovery of the benefits of computer automation and robot use is the reason why they have invested so much time and money in the research and continued development of these technologies.

Another factor that furthered Japanese interest and use was the oil crisis of 1973–1974 and its effects upon the world-wide economy. During this period, industrial production in Japan dropped over 14 percent and earned income to the work force dropped over 30 percent. This had a tremendous impact upon the Japanese economy, which also touched the U.S. economy. The drop in income caused the time-honored tradition of no layoffs in the work force to be severely tested during these times. The manufacturing industries absorbed these costs through losses posted on the books. Management developed strategies to cut into the losses by encouraging early retirement and cutting back on hiring. As the economic picture brightened in the late 1970s, Japan developed the strategy to implement more uses for computer technology in the manufacturing environment.

The major impetus was, however, the development of plans that had to do with the Japan's industrial work force itself. They were literally running out of workers. Japan has had a thinning population for decades, and the birth rate has dropped from 2.1 to 1.7 children per woman in the past decade. This fact, compounded with the reluctance of the young to enter the industrial work force (one that advocates seniority and rewards older workers with higher wages for performing the same job—almost twice as much for workers over age 40 compared to the wages for workers aged 20–24) painted a dark picture for continuing industrial growth.

In the early 1970s, the Japan External Trade Organization (JETRO) issued a report predicting that Japanese demographic trends, regarding the age of the work force and wages, would continue to rise during the 1980s. This was not good news; Japanese industry executives knew that the slanted wage scales for older workers would make attracting the young a tough proposition. Young Japanese, like young Americans, did not find the jobs in manufacturing particularly attractive as lifetime occupations. Part of the problem was that employment opportunities were shifting, as in the U.S., from the factory environment to other more promising careers, mostly in service work. The young people were also receiving more education. After graduation from Japanese universities, the young wanted higher-paying white-collar managerial jobs. Japanese industry faced a gradual decline of available workers, and they knew it. Computer automation and robots supplied the answer to the problem.

The acceptance of robots by the work force has been much more favorable in Japan than in the U.S., causing an increase in their implementation. As robots were used in more factories, workers generally accepted them quickly and found that they enjoyed the benefits the machines provided. The workers understood that the increased productivity helped not only the industry as a whole, but also benefited them as individuals. In spite of worker displacement (Japan has massive retraining programs for displaced workers), the industrial work force remained committed to the technology. As incredible as it may sound, robots, the simple machines that they are, are personified by the workers. Based upon many of the traditions of the Buddhist religion,

Japanese workers give the robots names and often speak to them, or of them, in kind, almost affectionate, terms. The Japanese worker tends to respond to automated robots, not so much as machines, but as close-to-human beings. American workers do not have such a kind attitude toward these machines.

Although the Japanese have been successful in introducing robots into the work force, Japan's labor unions seem to be changing their attitudes toward the technology. In place of total optimism, they are becoming more cautious in light of their economic problems and their much-publicized trade relations with the United States. Some unions in Japan have secured collective bargaining agreements that call for prior consultation with union representatives before a technological change occurs, total protection against layoffs due to robotic automation, no demotions or wage restructuring due to the machines, and more programs involving retraining and vertical education of displaced workers. Results of studies conducted by the Japan Economic Research Institute revealed that of the companies using robotics, most intended to use the skills of older workers after they had completed retraining programs. Their experience was simply too valuable to do away with. Hopefully, the U.S. will adopt this attitude as robot use becomes more common in our factories. Overall, the continued use and further implementation of robots will continue, due largely to the desire for improving the quality of the workplace and for the economic gains the automation facilitates.

The Japanese government also responded favorably to the use of automation and robots. Policies were developed calling for increased automation of manufacturing facilities, leading to more productivity, and countering the effects of the declining availability of workers. The U.S. government, on the other hand, does not recommend funding for robotic research and development, in spite of the success of the Japanese in this area. Japanese business leaders have a different attitude than does American manufacturing management; the Japanese philosophy is to sacrifice today for a better tomorrow. Japanese industrial management pursues long-term strategies almost in spite of the short-term costs. There is an even more striking dissimilarity between U.S and Japan that fosters Japan's automated advancements. In America, if a manufacturing company has a bad quarter, the news is in all of the papers, and the price of its stock inevitably suffers. Since U.S. manufacturing industries get working capital by issuing stock, a bad quarter can have disastrous effects. Everything a manufacturing company does is carefully scrutinized by its stockholders. If the value of a manufacturer's stock declines, it can quickly become a risky investment and will probably not be able to obtain additional money from other investors. This mentality can hinder the automation of the U.S. workplace. In Japan, 80 percent of manufacturing capital comes from banks, rather than from stockholders. Groups of companies associate themselves with banks to form symbiotic relationships. As a result, a factory in need of capital to purchase automated equipment has a much easier time obtaining the necessary funds. This relationship helps industry, and, at the same time, the banks grow economically.

The Japanese have also adopted programs that make the acquisition of robots much easier. Instead of requiring huge capital investments to buy these expensive machines, the Japan Development Bank, with the aid of the Japanese government, established the Japan Robot Leasing Company. This group was a cooperative effort of robot manufacturers and insurance companies that leased robots to large and small companies for a nominal charge, about $90 a month for a simple robot. When a company outgrew the robot or new technology replaced it, the company could trade up to the next model with only a small increase in the lease cost. This gave even the smallest Japanese factories the opportunity to use automation in their factories without making the large capital investment that would have been necessary in the U.S. This made even small factories in Japan more productive and efficient, because they could obtain the same automated equipment that the largest Japanese manufacturers were using. The Japanese government also provides special subsidies for industries buying robots and other devices for automating industry. They offer a 112.5 percent three-year depreciation of robotic equipment and provide many liberal breaks concerning export royalties. The Japanese government spends over $35 million a year in support of computer automation and robotic implementation and research, while the U.S. government spends less than $18 million.

The Japanese use of robots is most apparent in the automobile industry. Due to automated techniques, Toyota and Honda can produce cars that have an estimated cost savings in production of about $1,000 to $1,500 per car. Robots also increase the quality of the Japanese cars. Japanese automobile manufacturers proudly exhibit statistics regarding the reduced rate of repair for their cars in comparison to domestic autos. This further widens the cost advantage that automated production provides by the cost savings that are afforded due to the elimination of significant amounts of warranty work. Japan's robot-equipped Zama factory can build an automobile in nine hours, an amazing statistic. In 1980, the Japanese became the leader in world automobile production with an output of 11 million vehicles, exceeding U.S. production by an almost 40 percent. This is especially amazing, as the Japanese themselves do not see their automobile industry as an economic growth industry when compared to of bioengineering and computer electronics.

Why have the Japanese used automation and robotics so much while we have resisted? It is largely because of the differences in the way the Japanese conduct business. In Japan, the large corporations have policies that encourage cooperation between management and the work force. These policies include provisions for no layoffs, lifetime employment opportunities, lucrative profit sharing, and other incentive plans. Japanese manufacturing managers and corporate strategists are more interested in long-range goal achievment. This progressive outlook has resulted in a wide range of automated applications, some of which are straight from the pages of Buck Rogers—unmanned factories that work day and night without supervision and with little human involvement. The factory of Fujitsu Fanuc Ltd., the world's largest maker of CNC (computerized numerical control) systems for machining tools, is a technological wonder. It is a showcase featuring unmanned production techniques. During the day, factory workers produce robots and CNC machinery using automated techniques, while at night, the two-story factory becomes a high-tech wonderland. Computers control robots, machine tools, and conveyers in a totally automated environment building more robots. The only humans involved are those working in the computer room, monitoring the activities of the automated machines. In Nagoya, the Yamazaki Machinery Works, Ltd. also runs an automated factory at night. *The New York Times* carried this short description of the factory in a recent article: "What immediately catches the eye is the movement of the machines. They do not perform in unison, which is the characteristic pattern of traditional automation. Rather, each machine works independently, making an individual part different from its neighboring machining center. The computer tells a machine to drop one task, pick up another, speed up, slow down, or whatever—all in sync with the overall production plan."

The United States has nothing even near this level of automation, nor will we in the next five years, although our robots and computer technology are at the same level as our Japanese counterparts and, in some areas, more advanced. The Japanese feature heavy use of computer-integrated manufacturing (CIM). Robot production began to increase dramatically in 1976 when Japan implemented new strategies that were developed in response to the world oil crisis of 1973, and this technology has been growing ever since. Japan's use of computer automation and robots has simply made them more competitive. But, as secure as the Japanese manufacturing seems to be, they too, are being threatened in world markets. South Korea, with its low-paid work force and American investments, is challenging Japan's strangle hold on consumer electronic products. Robotics has little effect on this problem. As productive as robots may be, they still have a tough time competing with a low wage scale.

Japan employs more than four times the number of robots used by the United States. This is remarkable when you consider the respective size of their manufacturing economies and the actual uses for which they employ robots in manufacturing. The Japanese want to use robots in their manufacturing endeavors, while the U.S seems content to do research in robotics technology without widespread implementation of the automated machines.

While the U.S. develops robots from a stand-alone philosophy, not considering the integration of the machines into a real-world factory environment, the Japanese are busy developing robots for specific uses in their factories. For example, Kawasaki, Yaskawa, Hitachi, Nachi-Fujikoshi, and Mitsubishi, to name a few, develop robotics applications that help them today in their manufacturing plants. This real-world use of computer automation and its dividends is what separates the Japanese from the rest of the industrial world.

U.S. manufacturing companies have waited until the most sophisticated equipment became available, thinking that if it didn't have all of the bells and whistles it wasn't worth the investment. Japan, on the other hand, has implemented even the most simple automated robots without waiting and without negative results. In fact, the use of machines, such as pick-and-place robots, has resulted in a lowering of the cost curve in manufactured goods in Japan, causing a widening of the already large gap between more efficient and less efficient manufacturers.

The U.S. is actually better at developing high-tech products, but the Japanese have a better understanding of implementing this technology in the manufacturing environment. America has more advanced electronic gadgets, but we haven't used the technology to our advantage in the manufacturing environment. The Japanese have integrated the simple with the powerful, often integrating nonrobotic machines with computer-automated robots to achieve manufacturing efficiency. The Japanese use of robot automation evolves from their commitment to the manufacturing sector of their economy. They have developed and pursued a national goal of producing high-quality products. When capital investment decisions that include computer automation and robotics use are required, they are easily made when the considerations of high quality and savings in labor costs are factored into the equation. The Japanese know this, and it shows in their commitment to technology in their factories. Ominously, there is growing evidence that the Japanese, with their proven expertise at robotics application in manufacturing, are developing strategies to encroach upon the relatively untouched arena of factory automation and robotics in the United States. They profess that they want 25 percent of the U.S. robot market by the year 1990. Today, they account for approximately 9 to 10 percent. If this continues, the 25 percent mark may be a small percentage compared to the figure they actually attain. Unless they are impeded by superior products (which America already possesses) or further robotic implementation in U.S factories, the Japanese are likely to control and enjoy the lion's share of the computer-automation/robotics market world wide.

Fortunately, many U.S. robot manufacturers, as well as a large number of European companies, are realizing the strategy of the Japanese. These companies want to protect their investments and understand that they risk getting "burned" in the robotics marketplace unless they begin to form alliances with the Japanese. Naturally, the larger companies, like IBM, for example, can survive without such an alliance, but for many others, such alliances could be economically healthy. Some American companies have already entered into such alliances, but more need to do so. Until this begins to occur, Japanese vendors will continue to capitalize on their superior abilities in the area of factory automation and the concept of the computer-integrated factory, increasing further their competitive edge over the American (as well as the European) factory. In short, the economics of joint ventures cannot be understated. The comparative efforts of such cooperation in working and sharing technologies and implementation of ideas cannot be denied in the industrial world when the effects of not doing so are so severe to the companies themselves and the entire world economy.

CHAPTER
4

Design

Design is the first sector of manufacturing. Here, concepts and ideas become tangible realities. As simple as it may sound, design incorporates many different things. To produce a manufactured good, it is necessary to design three things: the product itself, the tools and machines that build the product, and the process flow of materials from machine to machine. This chapter covers basic considerations in the design of general robotics systems. Because robots are not used in the design of the product, and because design of the product is highly variable and specific, this chapter covers only those elements of computer-aided design methods that are used to design and program robot installations. The tooling required in the manufacture of robots is another product, with its own design phase. The process, unlike product design, is directly impacted by automation in the factory, and robots are a big part of that automation. Computer-aided manufacturing depends on the information that the computer gains about a specific process and by the ability of process equipment to react to inadequacies and compensate for them. Specific machines and processes may require change to adapt to the needs of computerized control systems. Such machines must be able to sense their surroundings, the output workpiece quality, and tool wear, in order to return this information to the computer. Because these abilities are expensive, the system must be flexible enough to produce many items and spread the equipment cost over all products.

Robot installations in tooling and process applications can benefit from the same computer-aided analysis that is used to design the product. Computer models of robots simulate the mechanical motions of welding, assembly, etc. to synchronize the robot with other machines. Complex variables, such as cycle times and torque-load vectors, can be analyzed to compare various robots. The computer models of all these operations require information about the abilities of the actual devices.

The basis of any manufacturing effort is the design drawing. Painstakingly prepared drafts and blueprints are used in

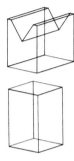

Figure 4.1
Some basic
geometric shapes
used in computer
modeling of solids

every step of a product's inception, design, and construction. But like the old-fashioned office, drafting is paperbound. Drawing is very labor intensive, and it is becoming a thing of the past. Unlike the electronic data in a computer, paper drawings cannot be manipulated, processed, and analyzed in fractions of seconds.

A computer-aided design (CAD) system is a tool used to automate design work. Instead of using a pencil and paper, the designer uses a sophisticated computer graphics display. A light pen is used to draw on the monitor screen or to extract lines, curves, and shapes from computer's memory. Solid shapes called from a computer library of geometric primitives can be assembled to form complex shapes, such as a robot. The basic three-dimensional blocks in Figure 4.1 are put together in larger modular shapes to define the Cincinnati Milacron T³ robot, the rotary table, and the workpiece in Figure 4.2. With a CAD system, there is no need to search through blueprint libraries or to transcribe what has been drawn before, no need to spend countless hours redrawing existing figures or parts assemblies. With a simple command, a designer can display a cutaway of a part, a closeup view, or an exploded view of all parts in an assembly. Even the structural foundation of a part can be displayed without the interference of viewing different layers of a drawing. Any product or part that was drafted on paper can be drafted using a CAD system. Some CAD systems are more complex than others, and the design capabilities are limited by the machine itself. However, even the most limited CAD system offers vast productivity gains over traditional manual methods of hand drawing.

CAD systems enable a designer to create complex computer models that can simulate the motion of synchronized pieces. As

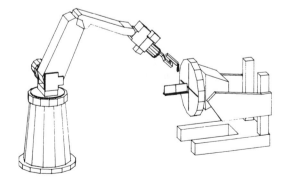

Figure 4.2 A complex shape made up of smaller fundamental shapes

shown in Figure 4.3, a car body is modeled next to a determinate structure model of a robot. This model helps determine if the robot can reach all necessary points on the car body as it moves down the line. The floor grid is used to show that both motions are relative to a third frame of reference. In Figure 4.4, the crucial aspects of a wheel well and the motion of the robot end effector are shown relative to each other. The robot motions developed from this model of the static relationship will be translated to the absolute motions required. The determinate model of the robot yields joint angles and motion parameters that are used to develop the programming for the actual robot.

General Electric has demonstrated its commitment to an all-out war on the productivity problem by the acquisitions of two well-known high-technology companies: Intersil Corporation and Calma Company. Intersil manufactures integrated circuits and cost GE over $250 million. Calma, a vendor of sophisticated CAD workstations, cost GE around $150 million. Together with an additional quarter-million dollars that GE intends to spend on research and development, this represents quite a sizeable investment. With all these new developments, the big hit at a recent GE press announcement was the news that GE has also bought the rights to use an assembly robot.

General Motors, a leader in automation technology, uses more than $100 million worth of CAD systems for original automobile design work. Using CAD, the more than 5,000 elements in a typical car assembly can be interrelated instantly, giving the designer the ability to add to the exterior framework or interior items. The computer images carry all physical dimensions of every item, allowing the designer to lay out a firewall, for example, using the actual clearances of the items that will be attached. Hose- and wire-routing schemes can be compared, with an exact materials takeoff produced for each scheme. Design items can be modified in seconds instead of hours, with the resulting labor savings being passed on to the customer through a lower price for the automobile.

This represents real savings in terms of the time it takes to generate a design, with direct impact on the amount of work and the number of designs that can be done in a given period of time by the design team. True productivity gains in the range of 200 to 500 percent are customarily realized within months of the installation of a CAD system.

With the incredible benefits of computer design, it looks like hand drawing will become as obsolete in manufacturing as using a slide rule for solving mathematical equations. But the control of drafting operations is only one part of a design automation strategy that affects every aspect of the design stage. A

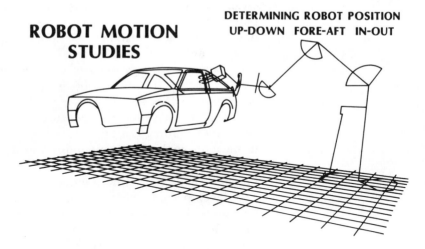

Figure 4.3 Geometric models of a car and a robot

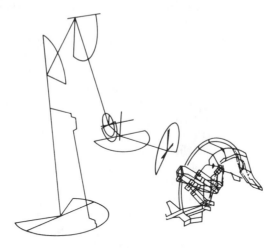

Figure 4.4 Geometric model of the robot performing work around an automobile wheel well

hidden impact is the identification of unworkable designs early in the process, eliminating expensive and discouraging difficulties in building physical prototypes. After the physical dimensions of an object are defined in the computer data base, mathematical models of the object can be built to simulate virtually any condition the part will encounter. The heat transfer capabilities of a piece can be compared for construction in aluminum, steel, and plastic. Complicated differential equations can be used to determine the mechanical stresses inside a strut or supporting frame, the forces and reactions at the joints, and the degradation and wear of a part over time. By easily performing such tests, computers change the actual design process, broadening it to include new methods and enabling one designer to do the work of a whole team.

Expanding the capacity of a single member of the design team is central to the idea of integrating separate processes, because it enables the team to cover functions that were previously allocated to an entire department. The designer not only takes over drafting but also has access to other disciplines in manufacturing that were previously viewed as completely separate. Through the computer, the designer can take manufacturability of the part into account, plan the production resources, and develop parameters for machine tooling. Removing the boundaries between each of these disciplines and condensing these functions into fewer people enables the design team to become the manufacturing team.

The design and product testing phases become closer, because the many aspects of a design can be tested without the need to build a working prototype. Many aspects of a product's performance in actual operation can be modeled on computers, negating or lessening the impact of road testing a car, for example. The data from a road test might, in the future, be derived from a computer-engineering simulation that takes place at the design stage. Engineers can use computers to simulate the dynamic mechanical testing of designs and analyze structural stresses in any product, making changes accordingly. In this way, the real-world behavior of a product can be analyzed long before the product is actually built.

Many applications of computers in testing are still experimental. With the exception of electronic printed circuits and microchips, testing through computer simulation is in its infancy. Thorough computer modeling requires that all factors affecting a design be taken into account before meaningful information can be derived. The more factors that are taken into account, the more accurate the computer model. Quantification of many factors is difficult. For example, a road test of a car driving down a bumpy road to determine if the interior is noisy requires that the contour of a bumpy road be digitized and a standard for noise be determined. These steps are not necessary in an actual road test, and their analysis shifts and expands the direction of the test effort.

Certain design jobs in different phases of manufacturing can be drawn together into the major design wing. Designs for the tooling, forging, and stamping machinery required to produce a part will come together with the design of assembly processes and materials-handling systems under the same roof as the design of the product itself. In this way, computer-aided design becomes computer-aided manufacturing.

Large numbers of numerically controlled machine tools already use central computers to control some operations. The next step is to integrate these isolated examples of automation into a comprehensive program. By implementing local area networks and establishing standard communications protocols, central computers can take advantage of the substantial data at their disposal to control operations in the factory. Computers are the tools by which engineers can reliably design and build complex items directly from a desktop computer. Many thousands of designs exist on magnetic media. With a totally automated factory, producing a prototype from any one of these designs becomes an exercise in data processing.

Computer-integrated manufacturing (CIM) covers virtually any use of computers in manufacturing. Numerically controlled machine tools are a simple example, although the term is most often used with complex integration. The numerical control program comes from a design data base that was developed on a CAD workstation and is treated like a construction blueprint by the CAM installation. A three-dimensional representation of a flange can be stored on magnetic media and transferred to a CAM utility, which will develop settings for machining or other fabrication processes. As part of the automated materials-handling process, a robot will pick raw billet metal or other materials and load them into the machine tools for processing. Taking information from the shop-floor robots and gauges, the manufacturing resource-planning (MRP) routines in the central computer control each step of the process. Any subassembly that requires work from several machines as part of a group technology can be considered a subprocess with its own quality guides and schedule. A flexible manufacturing system that includes several robots and machine tools can be easily (or automatically) reprogrammed to construct a different workpiece.

Consider a basic numerically controlled machine tool. Where does the numerical program come from? In the computer-aided design environment, a designer drafts the part to be constructed. The designer then calls upon computer-aided manufacturing programs to analyze the piece for manufacturability. The computer tests the software representation by simulating the geometric constraints of the milling bit as it traces out the shape of the workpiece. Thousands of possible trajectories for the bit can be compared in seconds. After the optimal process has been defined, the manufacturing resource-planning routines take over, scheduling the robots and machine tools of the shop floor and determining which manufacturing programs will be loaded into each tool and when. Other manufacturing resource-planning routines follow raw materials through the process, making sure that sufficient material is present at each stage and that the resulting products are moved to the next stage. Central to each of these processes are the robots that actually handle all the different pieces. Automated materials-handling robots insert raw workpieces into forges, presses, and machine tools, carry workpieces from one machine to the next, and assemble finished workpieces into complete devices.

Using CAD to Determine Motion

A continuous motion program can be developed from a dynamic computer model. A close-up of the robot and worktable in Figure 4.2 is shown in Figure 4.5. The junction of the two pieces of steel is the proposed path of the weld. The coordinates of the junction can be taken directly from the data base that was used to model the steel pieces. The trace of the welding tip is paral- lel to the junction, but the deterioration of the welding tip brings the end effector closer to the weld. To weld the other side of the junction, the robot must retract, so the table can rotate. View (a) shows the robotic welding tip approaching the weld location; view (b) shows an intermediate location of the welding tip during the weld.

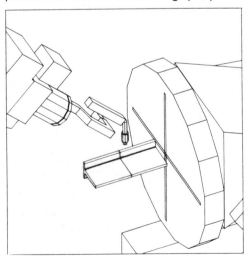 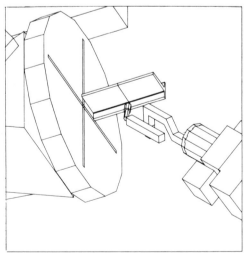

Figure 4.5 Modeling robot motion versus the workpiece

A complex system can retain all the economies of scale inherent in the large machinery in factories today and multiply cost savings by replacing retooling costs with changes in the various computer programs that control the processes. True flexibility in this type of system allows for the production of small batches of specialized items with no changes to the hardware involved. Batches of 100 or 200 pieces can be produced with all the advantages of dedicated machinery and no halt in production.

General Rules for the Design of Robotic Installations

To follow the automation trends set in the past, robot installations should start small. In a limited role, robots will not cure production problems, but if each subprocess is scrutinized independently, it is easier to integrate the subprocess with the existing system. You can't just plug in the robot and expect miracles, even with a small, easily defined task. If the production line is in trouble, rethinking of the total process is necessary. If the robot is supposed to solve a specific bottleneck or quality problem, analyze the whole process for hidden problems that might be compounded by the robot. The robot is programmed to do the same job as the existing systems, and, if the program suffers from the same defect that plagued the manual process, the robot will not be effective. New technologies are no place to troubleshoot production problems; solve as many problems as possible in the more familiar manual mode.

Designers of robotics installations must examine every facet of the investment: the robot, ancillary equipment, support and maintenance, and the factory environment. Failure to keep tabs on all aspects of the robot installation can lead to a myriad of unforeseen problems. Planning the installation of the equipment and the start-up operations leads to identifying those aspects of the factory that the robot impacts. Some adjustment of auxiliary functions, such as loading and unloading the machines that are upstream and downstream of the robot, will be required to make the best use of the robots. The robot will affect the entire process, and production planners must anticipate all effects and include them in the implementation schedules. Virtually all processes, no matter how small, must be subjected to scrutiny. Older methods may have to give way to new procedures. Although robots are made to serve people, some of the older methods may interfere with the plan to take full advantage of the robot.

No new technology can be successful without sufficient information and knowledge on the part of the people actually involved both in the new technology and in the application. Everyone involved must have a common aim and commitment.

Robotic Motion Dynamics

Different types of motion for a cylindrical robot are depicted in Figure 4.6. In point-to-point motion, only the angles and extensions needed to reach the starting and ending points are required. The arc is described by simultaneous extension and rotation and is more accidental than planned. A continuous path is identified by interim positions and angles. Each point is specified, but the path in between is still accidental. For controlled path motion, every point in a line is determined by a mathematical function, requiring synchronization of extension and rotation for the robot to pass through each point.

To follow a straight line, a jointed arm robot must rotate both joints at the same time, requiring two sets of dynamic vector calculations. These calculations and subsequent programmed instructions can be done by hand, but they can get very complicated when the trajectory is not a straight line (Figure 4.7). Even point-to-point operations can get complex, as shown in Figure 4.8. A continuous path can be programmed to change under certain conditions. A main routine for motion can include intermittent operations, or it can be set to branch. A sequence can be interrupted at any point and perform another set of motions.

Within each move, acceleration, constant speed, and deceleration affect the mass-torque relations differently (Figure 4.9). Most robotic moves have a minimum of constant speed time, because the moves

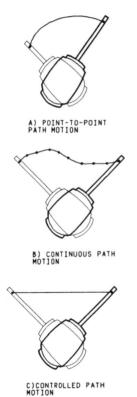

A) POINT-TO-POINT
PATH MOTION

B) CONTINUOUS PATH
MOTION

C) CONTROLLED PATH
MOTION

Figure 4.6 Point-to-point, continuous path, and controlled path robotic motion

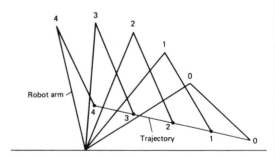

Figure 4.7 The trajectories of a robotic arm

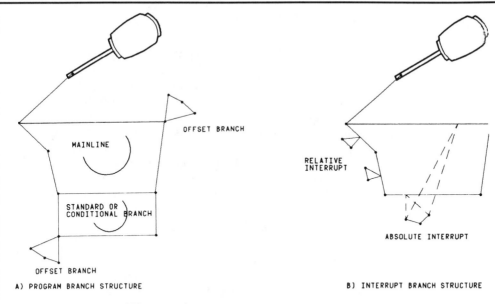

Figure 4.8 Branch and interrupt paths

CALCULATING OPERATIONS TIMES

Figure 4.9 Graph of robotic arm acceleration

do not reach maximum velocity and start decelerating halfway through the move. Different paths are required to operate on different workpieces. Figure 4.10 shows some examples of paths for a spot-welding gun. Straight-line moves can be achieved by motion along the major axes, but curved moves require attitude adjustment on the part of the wrist, or end effector.

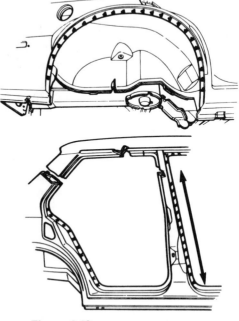

Figure 4.10
Examples of straight and curved motions

Upper-level management should voice concern over the success of initial robotics projects. This eliminates open opposition and stubbornness. The various teams involved must understand what is required and work together. Include all skilled personnel in the planning sessions for the installation. Their knowledge of the process will be an important factor in the final installation, and keeping them informed from the very beginning is a high priority. Be truthful about the possibilities for problems, the requirements of the robot, and the planned changes in work schedules.

The fundamental difference between conventional automation and a flexible robot system is the robot's ability to handle different tasks. If the application of a robot does not take this flexibility into account and use it to the utmost, some of the usefulness of robots is being ignored. Robots work at about the same pace as humans, but they never get tired, don't take breaks, and accurately repeat intricate movements. Present automation systems also have these advantages of consistency, repeatability, and endurance, but they are not flexible when it comes to producing different items. The key consideration in the initial installation of robots is how much flexibility is required now and in the future. How often will design changes occur? How many different designs are expected to be produced? If large numbers of tooling and machine changes are expected, robots can provide the flexibility required to minimize the effects of change.

Of primary interest in the initial applications of robots is the projected batch size and per-unit cost of the batch. An extremely low-cost item that is produced in the millions is not a good application for robots. The inherent flexibility of the robot is wasted when the robot does the same thing for long periods of time, and the robot's ability to do complex jobs is wasted when the workpiece is exceedingly simple. In this extreme, dedicated machines are the best solution. Items that must be produced quickly due to the sheer volume required are also better suited to production with dedicated machinery. At the other end of the spectrum is a unique workpiece that requires many hours of intricate working. If the robot is used to produce complex one-time items, the time taken to reprogram begins to dwarf actual production time. In these cases, humans are a more appropriate labor source.

The slated volume of production is also important to the cost justification of the robot system. When you buy a robot system, you are buying a 24-hour employee. Making sure that the robot is working all the time maximizes the investment. Conversely, a robot that is idle is comparable to paying an employee for sleeping at night. Most studies indicate that effective robot installations work the equivalent of two shifts per day.

Defining the working conditions and the nature of the work can help isolate potential applications for robots. Robots can work effectively in environments that would suffocate or otherwise injure people. Boring and frustrating tasks that would stifle humans cannot keep a robot from working at peak efficiency.

The return on investment in robotics systems can be misleading because the precedents established by fixed capital machinery are insufficient to gauge robot installations. After the robot is purchased, it becomes overhead. The equivalent is to pay a worker 50 years ahead of time on the promise of future work. The robot can't be laid off or fired. Robots have a useful life far beyond that of conventional machinery retooling schedules. The standard servocontrol motor can operate for more than 10 years before being rebuilt or replaced. The robot's ability to be reprogrammed makes each reprogramming the equivalent of a retooling effort. Avoid applying the financial guidelines of conventional machinery to robots.

Anticipate the robot's speed, production benefits, and the effect the robot will have on the upstream and downstream processes. Careful planning can eliminate the need for expensive parts banking and work-in-process inventories. Make sure that the robot will be kept busy and that the next stage in the process can handle the robot's output. Simpler standard machines can be rearranged to augment the robot's capacity. There may be time in the robot's cycle to include another task and take advantage of the robot's dexterity and speed.

The physical layout required for robot systems is different from that of conventional machinery. More open space is needed, because the robot moves through space at high speed. Barriers around the robot cell are necessary to protect the expensive machines that the robot serves, the operating personnel, and the robot itself. Entry to the work envelope of the robot must be restricted. The workspace does not need to be as dense as in the case of conventional machinery, because robots have greater ability to manipulate items and do not have bulky chassis. Future changes in the manufacturing process may require rearrangement of the robots and auxiliary equipment. This change in layout will be easier if permanent obstacles are kept to a minimum.

System layout plans should include as many of the machines and fixtures as possible. Not only does each motion have to be synchronized, but all machines in the work envelope present obstacles to the robot arm. Figure 4.11 shows a layout for a robot

SYSTEM PLANNING LAYOUT

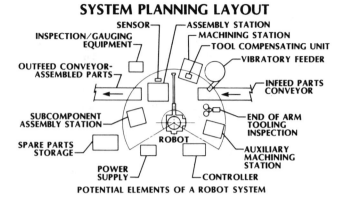

POTENTIAL ELEMENTS OF A ROBOT SYSTEM

Figure 4.11 An example of a layout for an assembly cell

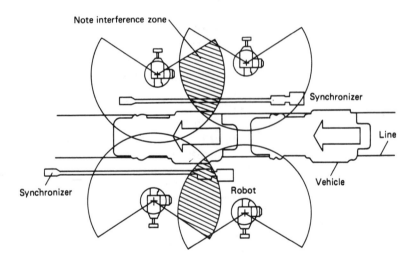

Figure 4.12 A multiple-robot station on an assembly line

that picks a workpiece from a conveyor, inserts it into a machine tool, gauges the work, and puts the finished piece on the output conveyor. It is easy to see that even in a simple example, the layout starts filling up in spite of the fact that the cabling and structural foundations are not shown. In multiple-robot installations, care must be taken to synchronize robot activities or separate the robots by enough space to prevent them from hitting each other. Figure 4.12 shows the interference zones for a four-robot installation. Different types of robots have different working volumes. Figure 4.13 shows the working volumes for each type.

Selection of the correct robot configuration is crucial, as might be expected. Robots seem to be able to do everything, and

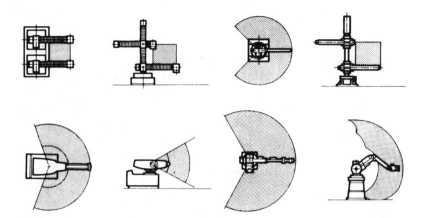

Figure 4.13 Different work envelopes of robots

this doesn't help the novice choose the right one. The various types of robots differ in subtle ways, and the options available for each type are mind-boggling. The correct choice is really a balance between the cost of the system and the current and projected needs. The robot must be able to accomplish the task at hand as well as a reasonable number of future tasks.

Planning should include a number of phases in the robotics build-up, with enough time for all people involved to learn from the initial system. Avoid the temptation to bring too many robots into a situation that has not stabilized. Don't make the mistake of thinking that one more robot will cure the problems. Additional robots and equipment can be purchased later, as the initial system becomes mature.

Robot functions can be divided into two main classes. Automated materials-handling robots present a workpiece to a machine tool, conveyor, or pallet. Continuous-motion robots maneuver a tool, welding torch, or paint sprayer about the workpiece. These are the fundamental concerns in any robot-to-application match-up. In handling workpieces, automated materials-handling robots must be able to deal with the entire range of weights and sizes. The loading capacity becomes a major factor. In welding and painting applications, controlled-path robots wield only one device of a fixed weight and size, and you can compute the loading on the arm. However, these applications require extreme accuracy, making dexterity, consistency, and speed of motion the main concerns.

Fundamental compromises must be made, using the following criteria for robot selection:

Repeatability	Dexterity
Consistency of motion	Weight capacity
Speed	Range of motion

Robots possess all of the above qualities in varying amounts, but an advantage in one area may be a drawback in another. A robot with a high weight capacity may not have the kind of dexterity required. You pay for these qualities, so don't buy more than you need. Decide which areas are critical to your application, and find the areas where you can make sacrifices.

Repeatability may be the most popular reason to use robots, and it is certainly a necessity for most applications. Most machine tool tolerances are measured in thousandths of an inch, and many operations are useless if this accuracy is not delivered. In some applications, however, for example, pallet loading, such accuracy may not be required. Methods for making the robot's job easier by allowing some flexibility (or error) in registering the workpiece on the pallet may prove to be cheaper than the corresponding increase in the robot's accuracy.

Welding, inspection, painting, and adhesive applying are examples of applications that require extremely consistent motion by the robot. Any jerks or twitches will be reflected in the finished job and may damage costly equipment as well. The tool must generally follow a contoured surface very closely, with little deviation from the path while performing a task along the entire surface. This operation must be done repeatedly with no deviation.

Speed of operation is at a premium if the robot must be able to keep up with the cycle times of auxiliary equipment. The timing of all these events should be carefully planned, so you don't push the robot's capacity. A robot moving its rated load at its rated speed should be able to perform the task, but there is a trade-off between the robot's speed, weight capacity, and reasonable maintenance intervals.

Welding and painting robots require a great degree of dexterity to reach some of the places that need welding or painting, but, as a general rule, robots are not required to perform excessive contortions. The more axes of motion a robot has, the more joints there are, which must be controlled in order to achieve a specific degree of accuracy and repeatability. Modern robots have astounding accuracy, but don't let the project's fate rest on the margin for error until you can decide for yourself that the robot can do the job over and over.

It is important that the robot be able to comfortably lift the largest tool or workpiece required for a given job, but there is a strong case to be made for positioning tables if the weight is excessive. Smaller (weaker) robots are often less expensive and usually have higher accuracy than larger robots.

The robot must be able to reach all the conveyors and machines in the workcell, but the robot's range must be consid-

Workspace Issues

Robot joint configurations must be analyzed to completely define the effective working space. Figure 4.14 shows the gross working envelope for a jointed arm robot and the effective envelope of the robot with an end effector. In the case of a welding robot, where the end effector is an extension on the arm, the work envelope may be expanded in some directions. In general, however, attitude moves of the gripper require arm motions that must be made inside the envelope, making the effective envelope smaller.

Figure 4.15 shows detailed parameters for the work envelope of a cylindrical robot. Points inside the work envelope may be accessible for several orientations of the arm. In the case of a cylindrical robot, these

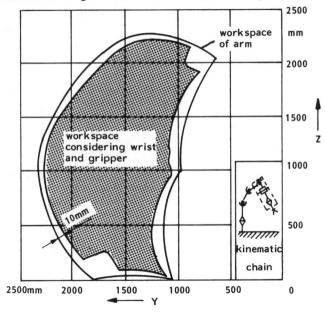

Figure 4.14
Workspace of the robot arm versus the workspace of the arm with wrist and gripper

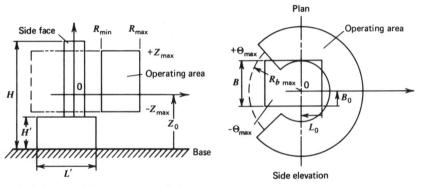

Figure 4.15 Workspace of humans versus robots

redundancies are few, while spherical robots and jointed arm robots have more possible orientations to reach the same spot.

When a robot is designed to replace a human operator, effective envelopes of the human and the robot should be compared. Figure 4.16 illustrates the comparison of human and robot work envelopes. Work envelopes should be carefully designed, taking into account all motions in the plan and elevation.

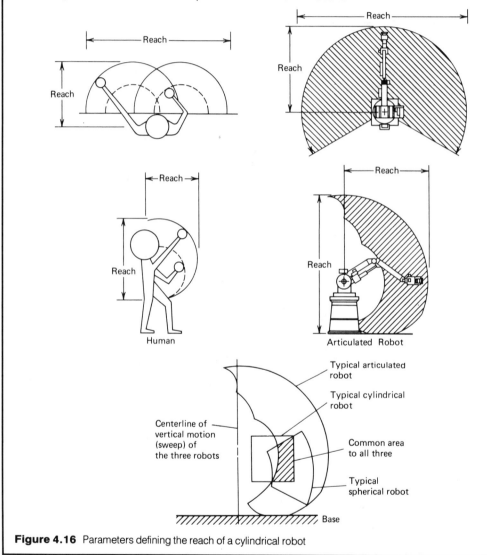

Figure 4.16 Parameters defining the reach of a cylindrical robot

ered along with the weight and speed required. Don't expect a robot to operate at its highest accuracy at the extremes of its reach. Range within some limit is perhaps the cheapest of the robot's inherent qualities, but extending beyond that limit may be prohibitively expensive. The range of a robot can be expanded in the future by mounting a pedestal robot on a gantry, for example. The work envelope of the stationary robot should be the maximum that is economically feasible, because inadequacies in range can make future applications unnecessarily difficult.

Robot accessories, end effectors, and grippers should be bought as part of the complete package. Most end effectors are highly specialized devices, and interfacing between the robot and the end effector can be a determining factor in deciding just how effective a robot installation is. There are no bargains to be found by shopping around and trying to integrate the system yourself.

Whether the robot is hydraulic, pneumatic, or electric, it must fit the application. Electric robots are the most precise. Electric motors have been built with incredible accuracies, taking advantage of the latest advances in feedback control systems. Several servomotors can be controlled exactly, making them the best choice for applications requiring complex, intricate operations. Electric robots are small, quiet, clean, and require the least preventive maintenance.

Pneumatic and hydraulic robots differ mainly in the working fluid used to translate motion. Pneumatic robots are best for light-duty materials-handling chores. They don't have the accuracy of electric robots but possess good range and speed. Hydraulic robots are best suited for heavy materials handling. They are the strongest, largest, and dirtiest robots. Due to the pressure of the working fluid, extreme care must be taken to maintain the various seals and bushings. A small defect could be disastrous, spraying fluid on the robot or workpiece and presenting a personnel hazard. Because they are the strongest robots, greater care must be taken to assure personnel safety. Hydraulic robots also consume the largest amounts of power.

Selection of the best control system is as important as selecting the right robot. The control program is the basis for the actions of all the robots in the system. As well as determining the ease of reprogramming, the control program determines the number of concurrent processes and the smoothness of the overall operation. Ease of reprogramming is essential in small batch operations and keeps retooling at a minimum. Unless there is sufficient funding to provide systems-level computer personnel, complicated control systems should be avoided. On the other hand, complex operations require complex programs. Make sure that personnel know the control system. The benefits of robots can be negated if the programming procedures are beyond the abilities of the floor operators.

End Effectors

Grippers are the most specialized of the end effectors. A selection of grippers is shown in Figure 4.17. Each gripper is specially designed to fit a certain application, and there is little overlap among these functions. In many cases, the robot arm (and the programming) will require changes to adapt to a different gripper.

Sometimes several pieces of equipment are needed to perform a task. Figure 4.18 shows a weld gun mounted with a clutch to absorb impact in case of inadvertent motion of the arm. The rigid bracket is used as an offset to provide added reach or a particular attitude adjustment, or to compensate for torque-mass considerations.

Everything past the robot mounting head should be considered an end effector, regardless of size or complexity.

Robots can be modeled on computers to provide analysis for bracket design. Figure 4.19 shows a robot arm and end effector. Given the desired attitude of the end effector, the robot arm finds the closest possible orientation. The bracket is designed to complete the link and account for any differences. The tool or end effector center point is taken into account as a displacement from the face of the robot wrist or mounting head. Figure 4.20 shows an analysis of the robot's tool center point.

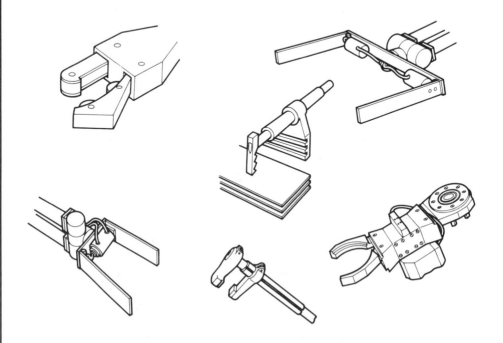

Figure 4.17 Typical end effectors

ASSOCIATED EQUIPMENT

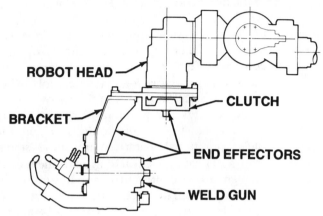

ROBOT HEAD

BRACKET

CLUTCH

END EFFECTORS

WELD GUN

Figure 4.18 Details of an end effector

BRACKET DESIGN FOR END EFFECTOR

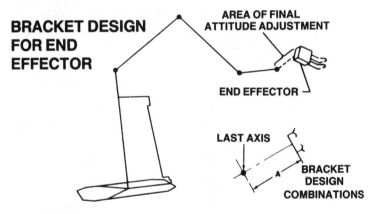

AREA OF FINAL ATTITUDE ADJUSTMENT

END EFFECTOR

LAST AXIS

BRACKET DESIGN COMBINATIONS

Figure 4.19 Robot model for the design of brackets

END EFFECTOR CENTER POINT

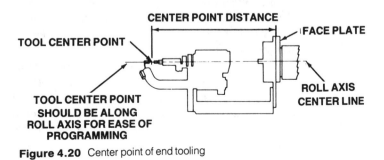

CENTER POINT DISTANCE

TOOL CENTER POINT

FACE PLATE

ROLL AXIS CENTER LINE

TOOL CENTER POINT SHOULD BE ALONG ROLL AXIS FOR EASE OF PROGRAMMING

Figure 4.20 Center point of end tooling

Programs should be easily changed, as required by changes in the fabrication process. Some aspects of CAM technology that are already in place can supply determining factors in the choice of a control system, but data in one format can usually be translated to another with little trouble. Many applications require a control system that can take information from sophisticated numerical programs for machine tools. The control program must control all functions in the robot workcell, synchronizing all movements and material flow as well as robot position and orientation. The system should have room for enhancements, such as additional operations or vision systems.

A pendant-style teaching instrument can be used to replace or augment the control program. In some cases, the program can be derived from instructions given to the robot from floor operators via the teach pendant. This has benefits not only in lessening the teaching time, but also can serve the floor operator as a means to get to know the robot. If the floor operator gives the instructions to the robot and actually walks it through its paces, then the floor operator is less likely to be surprised by the motions of the robot.

If possible, buy a complete integrated robotics system as a package. Robot vendors have seen and built robots for almost every type of installation. Make sure the vendor knows all the factors that concern your choice in a system. They can sell you all the hardware, software, accessories, and support that you need to have a fully integrated robot ready to be installed. In many cases, the vendor will have spent a lot of time investigating applications like yours, designing packages that require only a little tinkering to be finished. These investigations usually involve a person like you who has already made all the rookie mistakes possible, so take advantage of their experiences. But don't let anyone come in and tell you how to run your factory. You are the one who is going to have to live with the robot (and pay for it). Visit other robot installations similar to the one you have in mind. The robot vendors can put you in touch with people who can answer important questions about their installation. There is no substitute for experience, but your own experiences and observations will be the ultimate factor in the decision to automate further.

Keep your ear to the ground as to late-breaking advances in robotics. Things can change dramatically in this fast-paced field, putting expanded capabilities within your cost constraints. If it takes six months to plan, fund, and schedule your installation, make sure you don't miss out on a recent development that could really help you.

Sensing techniques are a good example of an area that will undergo extreme change. Welding systems use lasers to track seams, inspect welds, and determine start and stop points for the weld. Acoustic signals can be used like radar to give robots information about workpieces or obstacles in the workcell, as well as to test the quality of some types of welds and other processes. Vision systems eliminate the need for extreme accuracy in workpiece orientation or registration. Advances in tactility enable the robot to manipulate a workpiece more effectively. These advances in robot sensing have a definite payback in decreasing the costs associated with fixing the workpiece on worktables, pallets, and conveyors. They also allow greater flexibility in programming techniques and are necessary for adaptive control routines requiring information about process variables that cannot be supplied beforehand.

Specific Rules for the Design of Robotic Installations

Later chapters cover the constraints and requirements for robots in each of the main manufacturing areas: the fabrication, assembly, materials-handling, and inspection tasks. The remainder of this chapter presents specific information on all robotic systems.

The range, reach, and work envelope of the proposed robot must take into account the differences in performance that the robot can achieve at each point. Operations at the extremes of motion can be less accurate than operations in the middle of the robot's range. Different end tooling will alter the effective work envelope, and margins should be planned into crucial dimensions to compensate for this.

The load capacity of a robot is an important factor in the dynamic performance. In some cases, rigidity and stiffness of the arm, under full or partial load, is more important than gross lifting capacity. The exact movements of an assembly robot, for instance, must take the payload mass into account, but minimizing small deviations from the path will limit the mass as a function of required speed and acceleration. Load inertia, gravity, and oscillatory loading play a major role in selection of the end effector, because these forces will be applied at the gripper points. The overall load capacity is a function of robot arm acceleration, peak wrist torque, static structural deflection, steady-state motor torque, natural frequencies of the system, and the control system's feedback gain.

Robot manufacturers take the mass of the robot into account when planning moves, but the end effector, being specialized and adapted to fit a specific purpose, must be dealt with separately. Figure 4.21 shows a welding gun and other end effector pieces that must be considered. Welding cables are heavy, and counterbalancing equipment must be used to reduce the effects of the cables on robot motion. The respective centers of gravity for each end-effector component and the resultant center of gravity for the whole assembly are shown in Figure 4.22. All cabling, brackets, and tooling must be taken into account to calculate the motion and velocity of the arm.

The configuration of the robot joints and links is a function of the required motion, the control system, the obstacles, and the physical limits imposed on the structure of the robot. Because Cartesian robots move in straight lines perpendicular to each other, the control equations are easier to solve. If a straight path is required, Cartesian robots can achieve a specific gripper orientation more easily than robots with revolving joints. For a given work envelope, the Cartesian robot structures occupy more space than spherical or cylindrical robots, and although it is easier to coordinate two Cartesian robots, it is more difficult to overlap their work envelopes.

The degrees of freedom that a robot possesses determine its ability to place the end effector at an arbitrary point in the work envelope. Controlled path applications, such as spraying, require six degrees of freedom, but the trade-off for such dexterity is reduced payload and accuracy. Many tasks can be simplified to reduce the number of degrees of freedom required.

The gross work envelope of a cylindrical or spherical robot is determined by the range of travel in the major joints (shoulder

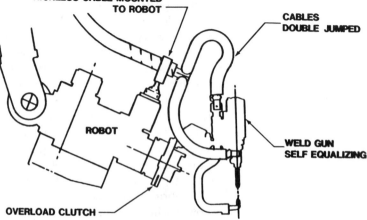

Figure 4.21 End effector centers of gravity

and/or elbow), with specific orientations of the end effector within the gross envelope determined by the wrist joint. Operating in confined spaces, the robot may require alternative joint configurations in order to reach a certain spot. The range of motion in each joint determines the number of such alternative configurations. Mating of threaded parts or coordination with rotating machinery may require a continuously rotating wrist. Controlled path trajectories may require 360 or 720 degrees of rotation.

There are four major components of a robotic joint; the joint axis structure, the mechanical power source, the transmission, and the position feedback device. Sampling rates for the feedback device vary with the structure; smaller structures have higher natural frequencies and require a higher sampling rate. A heavy joint or end effector incurs differences in inertial loading as the center of gravity moves, necessitating changes in the sample rate to compensate for changes in natural frequency and mechanical time constants.

The resolution of a robot is the smallest incremental movement the robot can achieve. The nominal resolution can be calculated from the resolutions of signal encoders and resolvers or motor step size, but the actual resolution is affected by friction, backlash, and the configuration of the robot. Typical resolutions range from .5 to .005 degrees for angular motion, and from 0.2 to .002 millimeters for straight-line motion.

TORQUE & WEIGHT CALCULATIONS

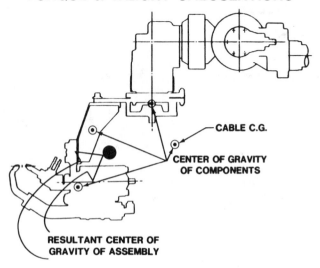

Figure 4.22 Mass considerations

Types of Robotic Installations

Robot configurations have different constraints and capabilities. Figure 4.23 shows four types of industrial installations. The jointed arm robot shown in the fixed floor installation is designed for applications that do not require large numbers of redundant orientations. The overhead gantry robot has perhaps the largest number of redundant orientations, because it can place the end effector at virtually any location around a workpiece and achieve various attitudes.

Figures 4.24 and 4.25 show two different robots at the same task. The floor pan for the automobile can be done effectively by the relatively simple Cartesian robot mounted on the floor, but the gantry robot, with its greater reach, can be programmed to work on other floor pans. The floor-mounted robot requires different pedestals to achieve vertical displacement, and it is also subject to more interference from the workpiece.

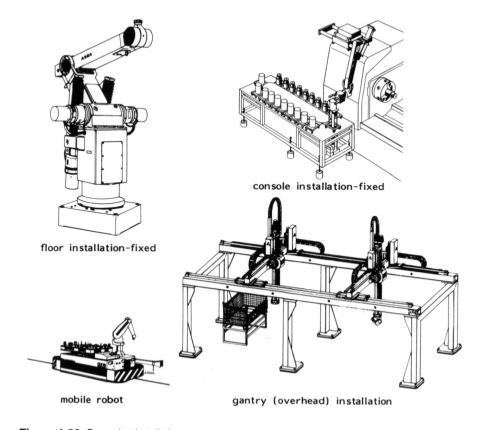

console installation-fixed

floor installation-fixed

mobile robot

gantry (overhead) installation

Figure 4.23 Four robot installations

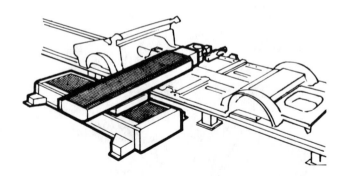

Figure 4.24 A floor-mounted two-axis Cartesian robot

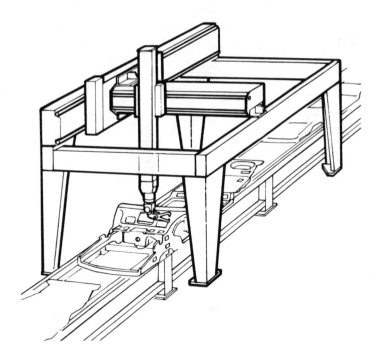

Figure 4.25 A gantry-mounted three-axis Cartesian robot

The requirements for the robot to be able to handle different workpieces can easily be met if a modular robot design is used. Figure 4.26 shows three robots that can be built from standard components. In the event of a new product design, the robots on the line can be added to or reconfigured.

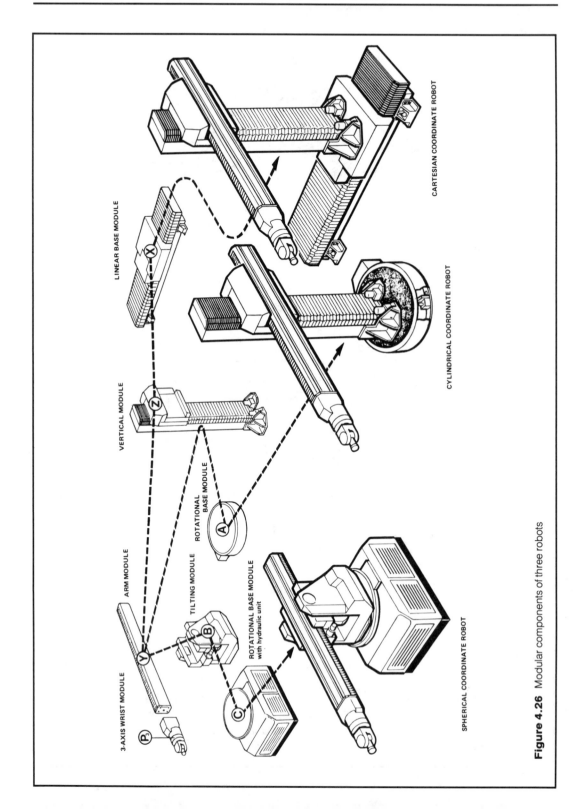

Figure 4.26 Modular components of three robots

Although they have a substantially larger supporting structure, gantry robots do not suffer from leverage and movement problems to the same extent as floor-mounted robots. Because they lift up, gantry robots use the tensile strength of their links. Figure 4.27 shows a gantry robot and its supporting structure. Table 4.1 shows typical payload and speed tradeoffs for such a configuration.

Figure 4.27 A gantry robot

PAYLOAD AND SPEED TRADE-OFF (DEAD LIFTING WITH 2 AXIS ONLY)

Z Axis		Load Capacity (lb.) Including Weight of End-Effector (Dead Vertical Lift)			
Tube Travel Maximum (in.)	Tube Speed Maximum (in./sec.)	3 Axis	4 Axis[a]	5 Axis[b]	6 Axis[c]
84	36	265	200	140	100
84	18	520	455	395	355
56	24	535	470	410	370
56	12	940	880	820	780
28	12	1430	1365	1305	1265
28	6	2200	2135	2075	2035

[a] Alpha and beta are limited to 7000 in./lb for one and two moving tubes.

[b] Alpha and beta are limited to 3500 in./lb for three moving tubes.

[c] Gamma is limited to 1000 in./lb independent of tubes.

Table 4.1 Payload and speed trade-off

Repeatability is described by the set of locations the end effector achieves when programmed to go to the same point repeatedly, carrying the same load and starting from the same point. This set of locations is usually described as a sphere of some nominal radius encompassing even the most erroneous placement. Standards for repeatability take system friction, joint backlash, servo gain, and mechanical clearances into account. Large robots for such tasks as spot welding attain specifications of +/− 2 millimeters, while precision micropositioners can achieve repeatabilities of up to +/− .005 millimeters.

Accuracy refers to the robot's ability to locate the end effector at a preprogrammed spot. The robot control system models the robot in relation to the world, taking arm deflection, joint type, and link lengths into account. The precision of this model and the models for tool and fixture determine the accuracy of placement. In many cases, a fixed registry is used to supply an offset value to compensate for accumulated errors. Accuracy is important in operations that follow a programmed data base, such as numerical control machining. Accuracies of up to +/− .01 millimeters can be reached with simple and accurate mathematical models.

The choices for robot implementation are complex and numerous. The computer-aided design methods used to design the product can be used to analyze the design of robotics tooling in the factory and to develop the programming of the robots. These methods also can be used to refine the process of fabricating and assembly.

Fabrication

After a product is designed, the next step in the manufacturing cycle, fabrication, begins. Fabrication takes us to the factory floor where machines cut and shape parts to make, for example, each one of the over 5,000 components of an automobile. Programmable automation to control these processes started decades ago. This chapter discusses the many applications of robots in fabrication and gives some basic engineering information on the implementation of robots in two of the major applications: spot welding and forging/casting.

When the initial forays into factory automation began, the machine instructions were stored on paper tape that had been encoded through punched holes, much like a music roll for a player piano. A machinist made the paper tape from design data that specified the position of the workpiece and controlled the operations of the different machines, table speed, and drilling or cutting depth. In modern factories, the paper tape has been replaced by a computer program that controls several machines and is much easier to alter than the punched paper rolls. The program is still written by machinists, but the machinists have been retrained in the use of numerical control programming languages. The machinists' extensive experience and knowledge of machining processes and tools is required to translate the design data into machine commands.

The John Deere tractor plant in Iowa is an outstanding example of what management that is willing to take a risk can accomplish with $500 million worth of computer-automated fabrication technology. Unfortunately, this automated factory floor is one of only 30 such installations in the U.S. In the John Deere plant, six programmable machining cells shape engine blocks. The cells do a multitude of different machining jobs, exchanging tools automatically, all without human intervention, under the control of computers. The system saves in set-up time, reduces scrap, cuts tooling time in half and work-in-process inventory by two-thirds, and the quality of finished parts is more consistent. Although the overall gain in productivity is hard to quantify, systems like these usually improve productivity by as much as 20 percent to 30 percent. Best of all, this system can quickly be changed to produce a different type of part. This flexibility is the big bonus of computer automation, because it enables a great variety of custom parts to be made at mass-production prices.

The essence of a sophisticated manufacturing system such as that of the John Deere plant is the flexible cell. The cell in Figure 5.1 has all the attributes required for flexibility: prioritization capabilities, multiple-parts handling, on-line inspection, buffer storage, input/output (such as magazines and pallets), and tool-changing. Figure 5.2 shows three views of a robotic tool-changing system. Placement of the tool rack must take the weight of each tool into account and not interfere with robot motion. Figures 5.3 and 5.4 show a layout for a flexible cell and a photo of the actual installation.

In the old days, when dedicated machinery was used to fabricate a piece, the whole system of machinery had to be replaced in order to fabricate a different piece. Completely replacing the mechanical fixtures and tools was extremely expensive and time-consuming. With computer controlled fabrication, changes in the fabrication process can be made simply by making a change in the computer program, eliminating the need for refixturing or retooling.

In the United States, only about 10 percent of all machining operations are computer controlled, but, already, computer-controlled machining accounts for as much as 50 percent of the value of all machining work. This directly follows from the higher productivity and higher quality of parts produced, and also from

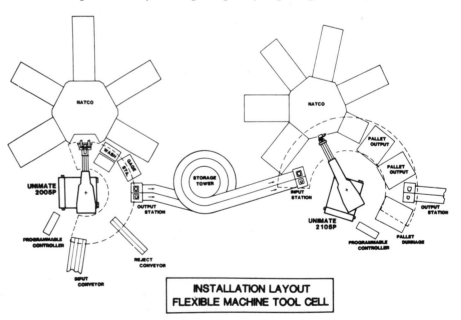

Figure 5.1 A typical manufacturing cell

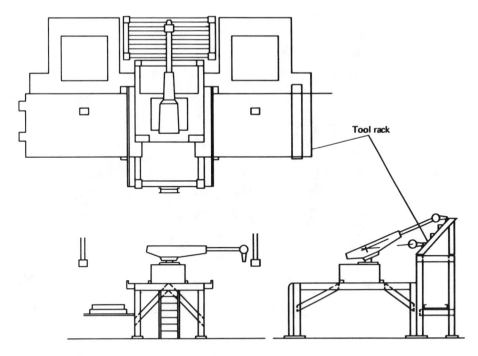

Figure 5.2 Robotic tool-changing system layout

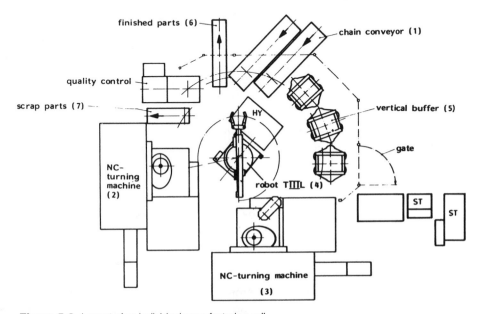

Figure 5.3 Layout of an individual manufacturing cell

Figure 5.4 An actual manufacturing cell in operation

the fact that fewer employees are required in automated machining operations. Although the direct employment effects cannot be completely known, many people are concerned about employment possibilities in the automated world. Many labor union leaders have adopted a negative, or at least skeptical, view of the roles that computer automation and robotic implementation will play in the employment picture of the future.

Many people feel that there will be a drastic change in the fabrication jobs that are held today by human workers. These jobs are going to become obsolete. In 50 years, today's factory will look as antiquated as the village blacksmith looks to us now. If cars were made with the same "time-honored" methods used to build horse-drawn buggies a hundred years ago, the Japanese, with their automated factories, or the Chinese, with their incredible labor force, would quickly put us all out of business. For good reason, many fabrication craftspeople and machinists already feel threatened about the prospect of losing their jobs.

The Bureau of Labor Statistics projects that there will be a shift in the next decade from high-paid manufacturing jobs to low-paid service jobs. Computer automation and robotics technology are expected to directly displace over four million manufacturing employees. It is projected that there will be around

100,000 industrial robots in service by the early 1990s. At a displacement rate of one robot to two people, roughly 200,000 jobs will be taken by robots. To combat the displacement, it is in the best interests of each employee to learn to operate the computers of automation. Along with this displacement, there is expected to be a gain of approximately 10 million service jobs. New jobs are also being created in computer and robot maintenance and programming.

Some large companies, General Electric, for example, have instituted retraining programs. This is more the exception than the rule, however. In the U.S., unlike our foreign competitors, there is no national strategy of government assistance for implementing automation in our factories. Many issues are decided independently and with no evaluation of the long-range effects and objectives. An extensive and serious dialogue between management and labor will be required to ease the many concerns regarding the introduction of computer automation and robotics. All layers of management and labor should begin consultations concurrent with the initial discussions concerning automation. Automation and robotics should be viewed as a necessary step towards successful competition. If the factory cannot compete effectively in the world market, it cannot supply jobs. Many studies indicate that one of the effects of increased factory automation will be an increase in total jobs rather than a decrease. As in the burgeoning computer market, the company will grow as a result of producing better-quality products at a lower cost.

Even so, computer automation and robotics have barely touched American industry. Many companies are not ready to make the capital investment that is required for conversion to complete automation. There are still 20 million factory workers and only 15,000 robots. Among the leading Fortune 1000 manufacturing companies, applied computer automation has found only about 5 to 10 percent of the estimated potential for growth and implementation.

An Overview of Fabrication and Processing Applications

The first applications of robots were in tasks with high degree of danger or discomfort for persons, for example, the noxious environments of spray painting and welding and the heat of the die-casting and foundry operations. Figure 5.5 shows an early industrial robot used as a die-cast tender. Although the robots were not necessarily economical, the elimination of dangerous jobs made them feasible. Later robots offered clear economic justification. Direct labor savings of 50 to 75 percent have been delivered by robot installations mainly because human

Figure 5.5 An early industrial robot in operation

labor rates have escalated,while robot costs have remained relatively constant. Increases in productivity are also gained by the robots' continual and slightly faster pace of work. Robots are also better where consistency and accurate repetition are considered more important than flexibility, dexterity, or judgment. In the future, the flexibility of sophisticated robots will be coupled with reduced costs, improved productivity, better quality, and elimination of hazardous jobs. The robots' ability to adapt to different products, design changes, and inconsistent workpieces will become an important cost justification.

Robot Capabilities

One basic operation on workpieces in the factory is transport from one location to another for machining, storing, assembly, or packaging. Robots are ideal for picking up an object and moving it to a machine or onto a conveyor. These types of operations can be done by relatively simple robots. More difficult tasks, such as sorting, machine loading, and palletizing/-packaging, require more sophisticated servo point-to-point robots.

Manipulation means twisting, inserting, or otherwise orienting the workpiece so that it can be worked on by machine tools, assembled, or processed in some other way. Robots can manipulate parts carefully, making them suitable for such tasks as machining, welding, assembly, and spray painting. Sophisticated manipulation can require a continuous-path or point-to-point robot with large program and numerical control data storage capacity.

Sensing gives a robot information that enables it to react to its environment. Sensor types include proximity switches, force sensors, and machine vision systems. The difficulty is in interfacing the robot with sensors and the capabilities and costs of sensing equipment.

Types of Applications

The capabilities of robots can be combined to provide a total capacity for specific applications. At the end of 1983, there were about 8,000 robots in the U.S., grouped loosely into seven categories:

1. Machine loading
   Diecast machines
   Automatic presses
   NC milling machines
   Lathes

2. Spraying
   Paint spraying
   Resin application
   Adhesive application

3. Welding
   Spot welding
   Arc welding

4. Machining
   Drilling
   Deburring
   Grinding
   Routing
   Cutting
   Forming

5. Assembly
   Mating parts
   Fastening

6. Materials handling
   Parts handling
   Palletizing
   Transporting
   Heat treating

7. Inspection
   Position control
   Tolerance

The automobile industry uses about 40 percent of the total number of robots installed in the U.S., with around 70 percent of these in welding tasks. Aerospace manufacturing also uses robots in a wide variety of tasks, as shown in Table 5.1. Although robots are used in many industries, about a third of the robots in the U.S. are installed in just 10 plants. Three applications—welding, materials handling, and machine loading—account for about 80 percent of the robots in the U.S. Even in welding applications, the market penetration of robots has been estimated at around 5 percent. This will increase as robots are improved.

Machine Loading and Unloading

Machine loading accounts for about 20 percent of the robots in the U.S. Applications include the loading of auto parts for grinding, loading auto components into test machines, loading gears on CNC lathes, orienting transmission parts onto transfer machines, loading hot-form presses, loading tranmission ring gears onto vertical lathes, loading electron-beam welders, loading cylinder heads onto transfer machines, loading a punch press, and loading die-casting machines. In addition to die-casting machines, robots are used for getting the workpiece from the conveyor, lifting it to a machine, orienting it, inserting or placing it on the machine, and transferring it to another machine or to a conveyor. One robot can service several machines or do other

Table 5.1 Representative robotized tasks in aerospace manufacturing

Kind of Task	Company	End Product	Description
Joining/Assembly			
Installing small parts	Boeing Aerospace Co.	Cruise missiles	Fastener feeding; installing cover fasteners
Welding	General Electric Company	Gas turbine engines	Fan frame hubs
Riveting	Fairchild-Republic Co.	A-10	Horizontal stabilizer
	Lockheed-Georgia Company	C-130	Bulkheads for floors and wings. Performed with stationary drill/riveter.
	Martin Marietta	Space shuttle	Portable drill/riveter is manipulated by robot arm.
Metal Working			
Chamfering	Pratt & Whitney Aircraft	Gas turbine engines	Vane slots in compressor shrouds
Deburring	Avco Aerostructures Div.	British Aerospace 146	Wing panels Panels up to 18 × 3 m (60 × 10 ft) possible using rail-mounted robot.
Drilling	General Dynamics, Ft. Worth Div.	F-16	Pilot holes in graphite/epoxy skin panels
	McDonnell Aircraft Co.	F-18	Canopies and windshield frames. Uses vision and five-axis DNC.

Table 5.1 Continued

Kind of Task	Company	End Product	Description
	Grumman Aerospace Corp.	F-14 and EA-6B	Fuselage panels. Rail-mounted robot serves four work stations and needs no templates to guide drill.
Machining	Martin Marietta Aerospace	Space shuttle	Removes 3.6 cm/sec (1.5 in./sec) from external tank ablator-covered parts. Uses tactile sensors to follow aerodynamic contours.
Routing	Grumman Aerospace Corp.	F-14 and EA-6B	Various sheet-metal parts. Rides along 6.1-m (20-ft) rails to serve four work stations.
Sanding	Boeing Aerospace Co.	Cruise missiles	Removes machine marks from wings.
Trimming	Boeing Aerospace Co.	Cruise missiles	Nose caps
Material Manipulation			
Composite layup	Northrop Corp.	F-5 and F-18	Vacuum lifting and stacking of precut plies.
Electronic circuit board assembly	Westinghouse	Avionics	Installs microcircuits onto circuit boards using two TV cameras.
Dipping	Pratt & Whitney Aircraft	Gas turbine engines	Dips wax patterns into ceramic slurries for investment casting.
Loading	General Electric Company	Gas turbine engines	Titanium slugs
Wire-harness routing	(Several)	Various aircraft	Gantry-mounted robots explored for electrical cables.
Applying Agents			
Applying sealant	Avco Aerostructures Div.	Fuel cells	Mixes and dispenses, allowing quality inspection at point of use.
Brazing	General Electric Co.	Gas turbine engines	Applies braze alloy to turbine nozzle supports.
Flame spraying	Garrett Turbine Engine Co.	Gas turbine engines	Metallic coatings
Plasma spraying	Pratt & Whitney Aircraft	Gas turbine engines	Ceramic and Zirconate protective coatings
Spray painting	Fairchild-Republic Co.	A-10	Mobile, rail-mounted robot; paints forward fuselage, nacelles, landing-gear tub, and other assemblies.
Other			
Inspection	General Electric Company	Gas turbine engines	Uses GE-developed light sensor to inspect 500 blades and vanes per hour.
	Pratt & Whitney Aircraft	Gas turbine engines	Positions turbine blades for radiographic inspection of cooling passages.

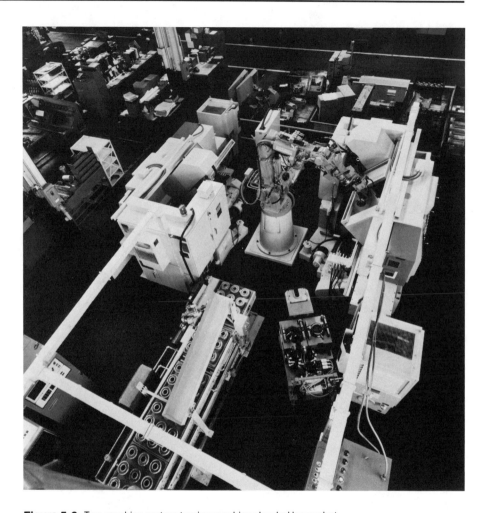

Figure 5.6   Two-machine system-turning machines loaded by a robot

operations while the machine is working. Robots are also used for the loading of hot billets in forging presses, loading machine tools, such as lathes and stamping presses, and loading plastic injection-molding machines.

An important distinction must be made between a robot loading a machine tool and actually performing the machining operation itself. There are many applications of the former but only a few of the latter. Figure 5.6 shows a robot loading turning machines. Figure 5.7 shows a robot loading workpieces into a machining center. Figure 5.8 illustrates the functional layout for the system, and Figure 5.9 shows a layout for one robot serving both a milling machine and a lathe. In Figure 5.10, another robot is serving two machines. Here, the robot is actually a simple pick-and-place robot taking workpieces and putting them on a rotary

Figure 5.7 A robot loading workpieces

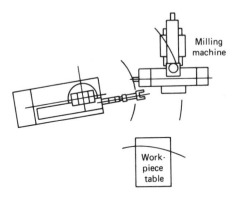

Figure 5.8 Layout of an application for a machining center

index table that cycles the piece through two machining centers. This is very similar to a materials-handling chore.

Spraying

Spraying and coating account for about 10 percent of the robots in the U.S. Applications include the painting of aircraft parts on an automated line, the painting of truck beds and the undersides of agricultural equipment, the application of the

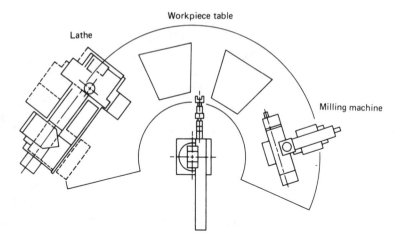

Figure 5.9 Layout for servicing two machine tools

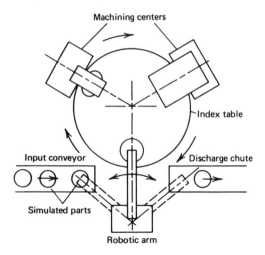

Figure 5.10 Example of a robot servicing two machines

primer coat to truck cabs, the application of thermal material to rockets, painting appliance components and car bodies, applying adhesives for sound insulation, applying paint, stain, or plastic powder to stationary or moving parts, automotive body panels, appliances, and furniture. Spraying robots use controlled paths, and the sequences of spraying can be coordinated with the motion of the conveyor.

Robots are also successfully used in time-sensitive thermosetting materials, such as the application of resin and chopped glass fiber to molds for producing glass-reinforced plastic parts and spraying epoxy resin between layers of graphite in the production of composite materials. Robots also apply adhesives to

Figure 5.11 A robot dispensing adhesive for sealing holes in refrigeration cases

various surfaces for holding of other parts, insulation, panels, etc. Robots can apply a more uniform coat than humans, reducing wasted coating material. They can also withstand the extremely toxic environments encountered in spraying the many new types of paints and coatings. Figure 5.11 shows a robot applying adhesives, and Figure 5.12 shows a robot applying sealant. Figure 5.13 illustrates a layout for a painting line. Five robots paint tractors as the line cycles them through washers, dryers, and baking ovens. The robots are carefully synchronized to prime and paint the whole tractor. Figure 5.14 shows a detailed elevation drawing of the robot in relation to the tractor chassis.

Welding

Spot welding is by far the largest application for industrial robots in the U.S., accounting for about 35 percent of the installed robots. Applications include the spot welding of auto bodies, welding front-end loader buckets, arc-welding hinge assemblies of agricultural equipment, braze-alloying of aircraft seams, arc-welding of tractor front-weight supports, and arc-welding of auto axles. Welding robots typically use point-to-point programming to maneuver a welding gun. Improved seam tracking systems enable the robot to perform the more sophisti-

Figure 5.12 Robot applying sealant to automobile body

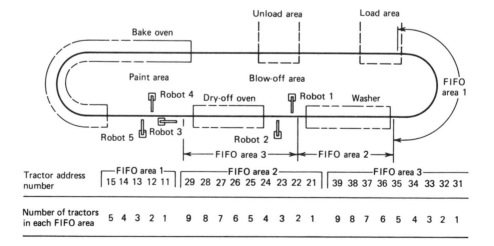

Figure 5.13 Diagram of robot-finishing system at John Deere Company's Tractor Works facility

cated arc-welding operations, as shown in Figure 5.15. The robot arc welding station in Figure 5.15 shows the barrier between the operator and the welder. As the robot welds the workpiece in the fixture, the operator places another workpiece in the alternate fixture. When the weld is completed, the two fixtures switch places. Robots weld more consistently, faster, and with higher quality than humans, and welding is hot, presents the known danger of eye problems, and exposes the operator to large electric currents.

Machining

Machining operations account for around 2 percent of the robots in the U.S. Applications include the drilling of aluminum panels on aircraft, metal flash removal from castings, and sanding missile wings. In most cases, the robot holds a power spindle and performs drilling, grinding, routing, or other similar operations on the workpiece. The applications are limited at present by the absolute accuracy of the robot and its sensing abilities. The workpiece can be registered by a human, another robot, or another

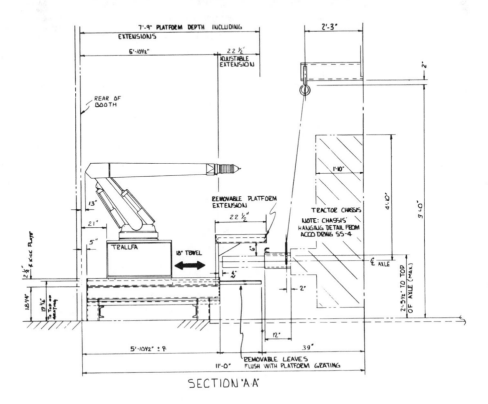

Figure 5.14 Diagram of robot positioning when coating tractor chassis

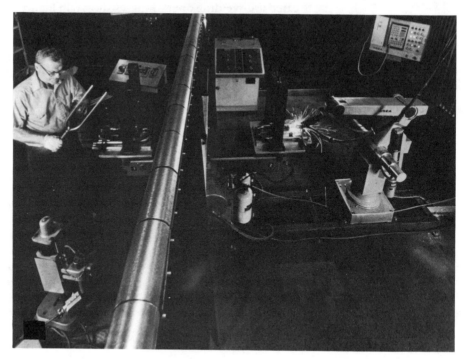

Figure 5.15 Robot arc-welding station with barrier

arm of the same robot, or the robot can move the workpiece around the tool, as in the case of a buffing wheel.

Robots can be used for limited tasks, such as drilling, routing, reaming, cutting, countersinking, broaching, and deburring. These difficult machining operations test the abilities of the most sophisticated robots at present. The major difficulty is positioning accuracy. The accuracies in machining operations are commonly measured in thousandths of an inch, forcing robots to rely on jigs and fixtures that eliminate some of the cost savings and flexibility advantages of robots. Increased accuracy and adaptive control using sensory equipment are needed for robots to successfully compete in these operations. Some machining operations are very involved, requiring complex programmed paths and high speeds that the robot cannot deliver.

Figure 5.16 illustrates the basic task of machining operations, making a product from raw workpieces. Figures 5.17 through 5.19 show a robot applying the workpiece to a lathe after removing the blank from the feed station. Here, the robot takes an active role in actually performing the machining operation. Figure 5.20 shows the layout corresponding to the lathe and robot shown in Figure 5.19.

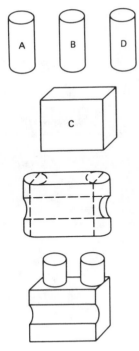

Figure 5.17 Robot removes blank from feed station

Figure 5.16
The two objects in the bottom row are obtained from machining operations on the primitive volumes (A, B, C, D).

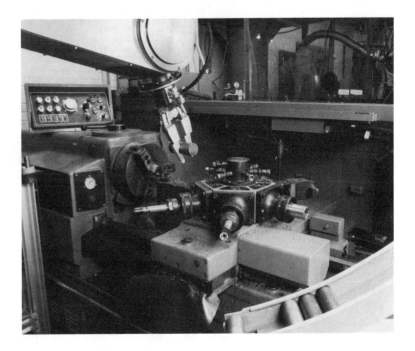

Figure 5.18 Robot carries blank to lathe

Figure 5.19 Overview of an application to a lathe

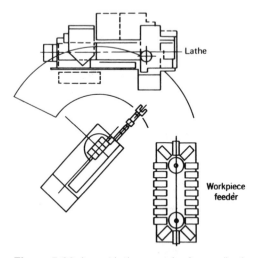

Figure 5.20 Layout in the example of an application to a lathe

Assembly

Assembly accounts for about 3 percent of the installed robots in the U.S. Applications include the assembly of aircraft parts with auto-riveting equipment, riveting small assemblies, drilling and fastening metal panels, assembling appliance switches, and inserting and fastening screws. Robots also are used to insert light bulbs into instrument panels, to manufacture printed wiring boards, to assemble typewriter ribbon cartridges, and to build small electric motors. These applications are also limited by the robot's accuracy, programming, and sensing. Chapter 6 of this book details robots in assembly tasks.

Materials-Handling

Automated materials-handling (AMH) applications account for about 25 percent of the robots in the U.S. Applications include the moving of parts from warehouse to machines, depalletizing wheel spindles onto conveyors, transporting explosive devices, packaging toaster ovens, stacking engine parts, transferring automobile parts from machine to overhead conveyor, transferring turbine parts from one conveyor to another, loading transmission cases from roller conveyor to monorail, transferring finished auto engines from assembly to hot test, processing thermometers, bottle loading, transferring glass from rack to cutting line, and shelf making. The transport capabilities of the robot are more important in materials handling than its pure manipulative abilities. The motions can usually be defined in two dimensions but can include three. The robot, typically a nonservo or pick-and-place robot, can be mounted on a stationary floor pedestal or on rails that enable it to move from one workstation to another. Robots are also used for the transfer of parts from one conveyor to another, the transfer of parts from a processing line to a conveyor, palletizing parts, and loading bins and fixtures for subsequent processing.

The benefits of robots in materials handling are reduced direct labor costs and removing humans from tasks that may be tedious, exhausting, or dangerous. There is also less damage to parts during handling, especially with fragile objects. Robots are generally not applicable to materials-handling chores for large production volumes or where no manipulation is needed; fixed machinery is better suited in these cases. Chapter 8 deals more specifically with robots in materials-handling applications.

Inspection

Inspection applications for robots are just beginning and only a neglible percentage of the installed robots are used in this area. Some robots are used in inspecting dimensions of parts and inspection of hole diameters and wall thicknesses. Inspection robots are used with sensors, such as television cameras, lasers, or ultrasonic waves to check part location, identify defects, and recogize parts for sorting. These applications are strongly limited by sensing abilities, programming techniques, and robot accuracy.

Casting and Forging Applications

Die Casting

The second largest application for robots is die casting. Robots were first used in this application in 1961. The General Motors die-cast machine tender was ideal, because die-cast parts are produced in large numbers, but the time for equipment changeover must be kept to a minimum. Parts are already oriented when they are removed from the cast, which allows robots to pick them up with standard grippers. The die-casting process has been around for a long time, minimizing the equipment- and product-design changes that necessitate retooling.

Robots can be used in removing the part from the die, quenching, trim-press loading, and periodic die maintenance, insert placement, and conveyor loading. A single robot can tend two die-cast machines, depending on casting cycle times, physical layout, and robot speed and type. Interfaces with other machines can be complex and sensory equipment may be needed to assure that all stages are completed. Two workers per shift can be replaced by one robot, making a savings in direct labor cost of up to 80 percent. Because of added consistency, throughput increases of 20 percent are possible. Significant increases in quality can be gained because of constant die temperatures, and net yield increases of 15 percent have been reported. These factors go together to increase die life, reduce the required floor space, and decrease the cost of safety equipment, because humans are not present.

Figure 5.21 shows a robot serving two die-cast machines. The robot removes the castings, puts them in the quench, and places the pieces on the output conveyor. Figure 5.22 shows a similar robot that removes the casting, quenches it, inserts it into a trimming machine, and onto the conveyor.

Foundry Operations

Foundries present an environment and job that are very difficult for humans, but robots still have not penetrated well into this area. An increase of robot use is expected, even though foundries have a large number of diversified castings, and a low-tech approach is usually used. Applications include mold preparation, ladling of molten metal into molds, final cleaning of castings, core handling, spraying and baking of refactory washes on copes and drags, and casting removal from shakeout conveyors. The many coating and drying cycles of investment castings are particularly suited for robots. Investment casting highlights the consistency gained through robots, because uni-

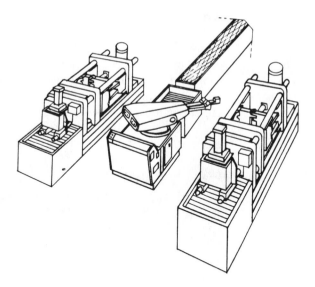

Figure 5.21 Robot unload of two die-casting machines with quench

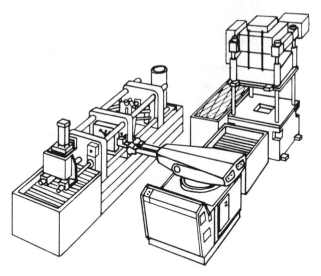

Figure 5.22 Robot unload, quench, and trimming operation with parts removal to conveyor

form molds are crucial to quality. The problems for robots are the abrasive dust, variations in processes, and the sensing equipment needed to clean molds and castings.

Forging

Robots in forging are used primarily to load large presses and furnaces, move heated billets from the furnace to the drop hammer, forge presses, and move forged pieces from the press to conveyors or pallets. Robots can be used in virtually all phases of forging, showing throughput increases of up to 30 percent, but several problems remain. Handling of hot pieces requires special equipment or periodic cooling of the end effector, and the heavy shocks encountered by the robot require isolating the grippers. The robot must compensate for time-varying warpage and die wear, and a high level of interfacing with other equipment is required. Figure 5.23 illustrates the use of robots in forging.

Heat Treatment

This application primarily involves materials handling and loading of heat-treating furnaces, salt baths, and drying apparatus. The robots provide more consistent cycle times, cost reduction, and increased quality while eliminating hazardous jobs. As shown in Figure 5.24, the ability of the robot to withstand high temperatures comes in handy for such operations as laser case hardening of metal pieces. Cladding and surface alloying also can be done with minor changes to this station.

Forming and Stamping

Low-technology pick-and-place robots are used to feed small presses and load large-body stamping presses. Robots provide lower costs and eliminate a dangerous job. Robots cannot compete with the large, fast, dedicated machinery used in very large production runs for stamping, pressing, and forming. Figure 5.25 shows a robot inserting a workpiece first into a furnace and then into a die-forming machine.

Plastics Processing

Applications include loading injection molding machines, transfer molding presses, structural foam molding machines, handling large compression molded parts, and loading inserts into molds, applying resins and secondary trimming, drilling, buffing, packaging, and palletizing of finished products. About 5 percent of injection-molded machines are robotized, with Japan leading the field. The robot can be mounted on the injection-mold machine, or it can be mobile to tend several machines. Low-cost application-specific robots have been developed especially for this purpose. The major reason for using robots is to achieve cost savings due to both the reduction of labor costs and increased cycle throughput. Quality is increased through more uniform process cycles and more consistent handling of delicate

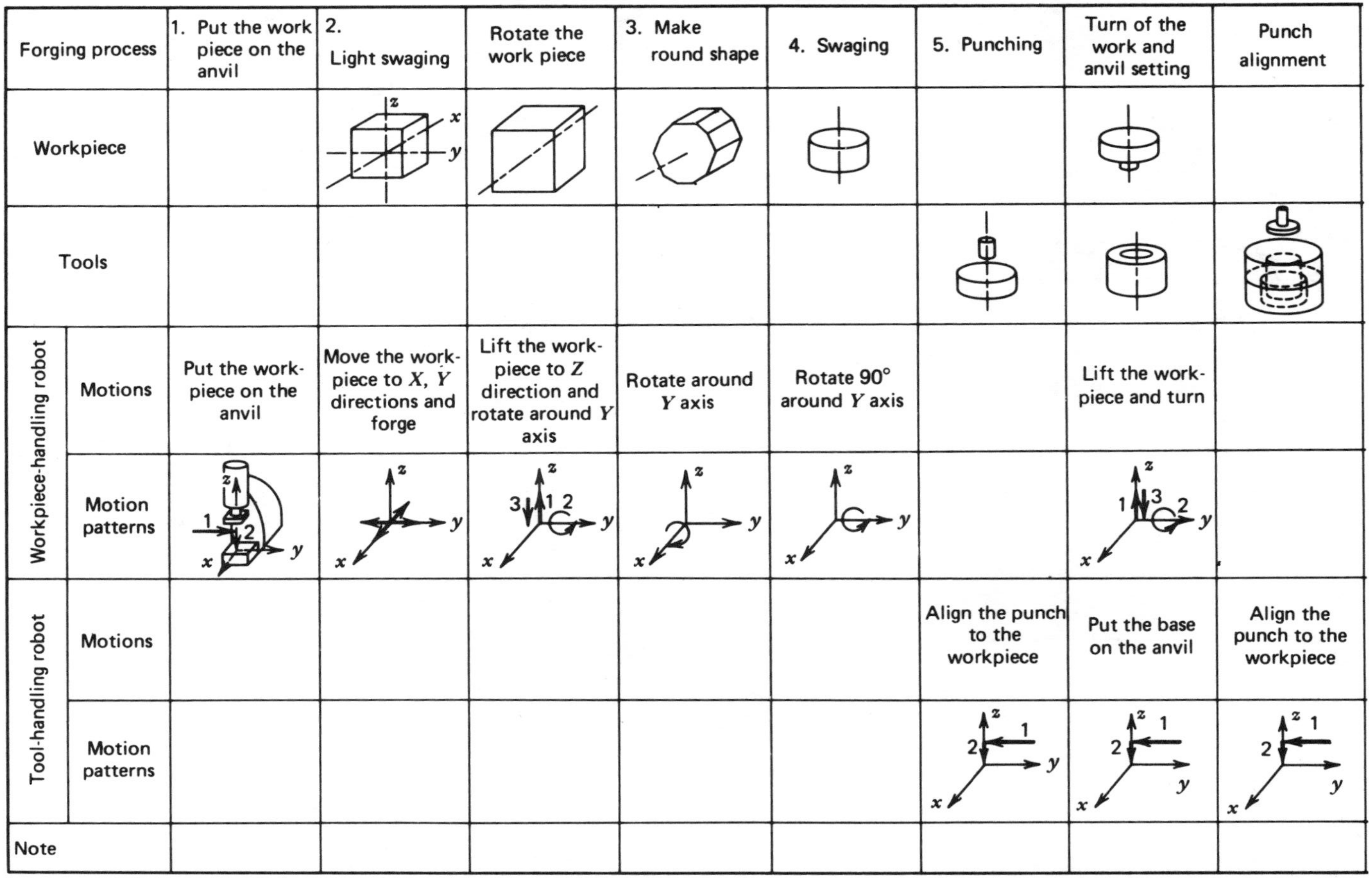

Forging process	1. Put the work piece on the anvil	2. Light swaging	Rotate the work piece	3. Make round shape	4. Swaging	5. Punching	Turn of the work and anvil setting	Punch alignment
Workpiece								
Tools								
Workpiece-handling robot — Motions	Put the work-piece on the anvil	Move the work-piece to X, Y directions and forge	Lift the work-piece to Z direction and rotate around Y axis	Rotate around Y axis	Rotate 90° around Y axis		Lift the work-piece and turn	
Workpiece-handling robot — Motion patterns								
Tool-handling robot — Motions						Align the punch to the workpiece	Put the base on the anvil	Align the punch to the workpiece
Tool-handling robot — Motion patterns								
Note								

Figure 5.23 Details of motions and motion patterns in robotic free forging

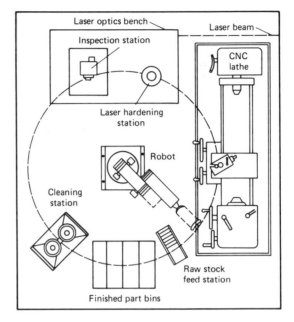

Figure 5.24 Laser heat treatment layout

pieces. The robots must interface with other machines, such as trim presses and conveyors, and the gripper must be specially designed because of workpiece size, quantity and condition.

Electronics Processing

This new field for robots includes machine loading, component placement, microcircuit assembly, component insertion, printed circuit board assembly, cable harness fabrication, and robot-assisted testing and inspection.

Glassmaking

Robots are used in glassmaking because of their ability to withstand high temperatures and to handle fragile workpieces. Robots have been used to charge molds with molten glass and for handling both sheet and contoured glass.

Primary Metals

Materials-handling robots have been used to charge and tap furnaces.

Textiles and Clothing

The limp nature of the workpieces challenges robots, but they have been used to handle material in spinning operations and to place and sew clothing items.

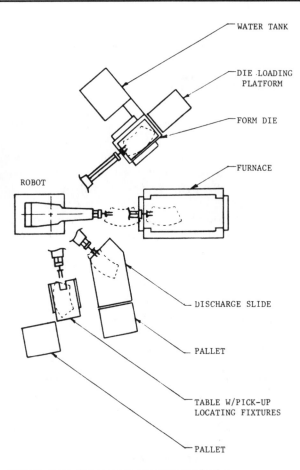

WATER TANK

DIE ·LOADING
PLATFORM

FORM DIE

FURNACE

ROBOT

DISCHARGE SLIDE

PALLET

TABLE W/PICK-UP
LOCATING FIXTURES

PALLET

Figure 5.25 Robot in press-forming application

Food Processing

Materials handling and packaging of food are two applications, along with the actual processsing of foodstuffs and decorating candies and chocolates.

Chemical Processing

Although a continuous activity, chemical-processing applications of robots include cleanup and maintenance of equipment.

Welding Applications

Spot welding did not become the most common use for robots because the operation was simple. Spot welding is a complex task, but because of the need for increased accuracy, speed, and reliability in welding large numbers of sheet metal items, the cost benefits of robotic welders led to their wide-spread use. Advances in robotic welding are due in large part to the activities of major manufacturers such as General Motors and General Electric.

In 1969, GM installed 26 spot-welding robots made by Unimate. In 1970, Mercedes-Benz in Europe installed similar robots. Since then, use of robotic welders has grown to account for 25 percent of the robots in Japan and 30 percent in the U.S. The repeatability and positional accuracy of robots, together with greater control of individual welds, provides greater throughput and quality than manual methods of welding.

Welding is the process of joining metals by fusing them together. Soldering and brazing, on the other hand, join metals by adhesion. In spot welding, sheet metal sections are joined by a series of joint locations, or spots, where heat generated by a large electric current causes the sheets to fuse together. The gap between the sheets, the amount of current, and the duration of the current pulse are the three most critical factors in the quality of the weld. In most cases, the gap between the sheets is controlled by the electrodes themselves applying pressure to the sheets right at the location of the spot. Figure 5.26 shows the electrodes of the welding gun and the sheets to be welded. As shown in Figure 5.27, the resistance of the complete secondary circuit is a function of the secondary winding, cabling, electrodes, top and bottom sheets, and the gap between the sheets. If the gap is too small, the resistance is reduced, and higher current is required to make the weld. If the gap is too large, resistance is higher, and the spot can burn through. The gap, current, and time must be adjusted to be optimal for different material thicknesses and compositions.

In a typical welding sequence, the electrodes grip the sheets to allow the required gap and apply the current for the specified time, generating heat that melts a column of metal between the

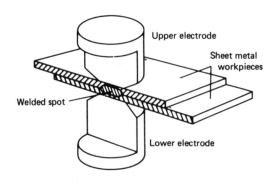

Figure 5.26 Electrode and worksheet placement

electrodes and filling the gap. The electrodes hold the pressure on the sheets until the spot has cooled. Some welding electrodes are internally cooled. The electrodes release the sheets and wait for the next workpiece. The automatic control of robot and welder can repeat this sequence accurately and adjust for different welding conditions.

The spot-welding robot is composed of three main sections: the mechanical assembly of the body, arm, and wrist of the robot, the welding tool, and the control unit. The robot body is used to position the welding tool in the working volume and to orient the tool with respect to the workpiece. The criteria for selection of a robot to fill this purpose are as follows:

1. The number of degrees of freedom
2. The maximum stroke along each axis, including electrodes
3. The total volume of movement, including electrodes
4. The maximum load at maximum speed for any given path (40–100 kilograms at 0.5 to 1.5 meters per second or 120 to 240 newton-meters at 60 to 80 degrees per second)
5. Position accuracy (0.5 to 1.0 millimeters)
6. Repeatability (1.5 to 2.0 millimeters)

The resistance-welding tool, or gun, is composed of a transformer, a secondary circuit, and a pressure element (a gripper-type electrode). When the spot distribution is straight, and there are no access problems, up to 60 welds per minute can be done. To achieve this rate while welding two 1-millimeter sheets together requires a 10-kiloampere current pulse for 10 cycles. The force applied to the sheets by the electrodes should be in the range of 3000 to 3500 newtons.

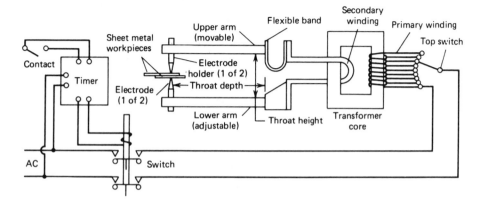

Figure 5.27 Electrical components of a welding gun, showing the transformer and secondary circuit

Problems with access determine the use of three types of robotic welders; those with an overhead transformer, an on-board transformer, or a built-in transformer. Figure 5.28 shows these basic configurations. Overhead transformers are usually suspended on a track above the robots and move independently to follow the movements of the robot arm. The secondary circuit cable must be long enough to absorb all motion of the arm and rotations of the wrist. An independent balancing device is usually used to bear some of the weight of the cable. On-board transformers are mounted on the robot as close as possible to the welding gun. This shortens the secondary cable but increases the weight the robot must move, thereby slowing the operation. Built-in transformers are mounted right at the end of the arm, eliminating the secondary cables. Because the secondary resistance is

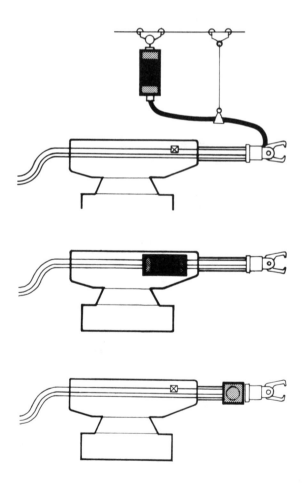

Figure 5.28 Three transformer-mounting arrangements

reduced, the transformer does not need to be so large. This solution makes the welding gun bulkier and heavier, posing access and robot-strength problems. Robots in this configuration must be able to handle about 200 newton-meters.

The control system must include programmability and process control in addition to actuator control. Actuator control refers to the control of the robot, welding gun, and workpiece. Programming leads to the kind of flexibility required by an expensive manufacturing installation. Process control involves using an AC phase-shift controller and counting the number of current periods, or cycles, for a particular weld time. This can be done by using a timer, but a more effective and flexible method is to integrate process control into a central controller. Central controllers keep maintenance and interfacing problems to a minimum by integrating materials handling with the robot positioning, adaptive parameter information acquisition, and communications. Figure 5.29 shows two modular welding robots, (a) a scissors-type gun and (b) a C-type gun. Both are modular, have integrated transformers, an electric current-voltage rating of 33 kilovolt-amperes and a force rating of 3.5 kilonewtons.

The quality of individual welds can be tested by the dynamic resistance method to identify welds that must be done over and to monitor conditions that would cause the welding electrodes to weld themselves to the workpiece. If an inadaquate weld is found, robots down the line can be automatically programmed to put new welds around the faulty weld. Other process problems, such as a malfunctioning gun or improper settings, can also be identified. If the electrodes stick to the workpiece, the controller must interrupt the welding sequence to prevent the robot from moving to the next weld position, and it must also signal maintenance crews.

Central control of both the process and robot is essential to the development of adaptive programs, making it easier to accommodate changes in weld patterns and workpiece variations for various models. Combining overall line control with process control allows for power interlocks so that current is allocated to each gun independently, maximizing power resources. Maintenance factors, such as electrode wear, can be monitored constantly, with appropriate changes in weld current and force adjusted automatically.

Criteria for spot-welding facilities are as follows:

1. The parts to be welded
2. The geometry of the parts and the number of stations required

Figure 5.29 Modular 33-kVA, 3.5-kN welding gun with integrated transformers. Both have four different mounting surfaces for adaptation to the piece-part geometry and self-equalizing secondary circuits; (a) scissors-type gun (b) C-type gun.

3. The distribution of the welds
4. The production rate and the number of lines
5. The desired degree of flexibility
6. Transfer and positioning of the workpiece
7. The final selection of the robot, equipment, and installation
8. The available space and environment

Before you set up a spot-welding system, make a detailed study of the parts involved and the assembly process. Classify the parts into three rough groups according to their size and sequence in the assembly process. In the first group are small pieces, such as tabs or supports, that are welded onto large pieces. Robot access may be a problem when welding small pieces onto the center of a large piece. Next, there are pieces of about the same size that are welded together to form subassemblies. The third group contains the subassemblies that fit together to complete the form to the main structure.

The geometric references are the significant zones of a part that define its theoretical position relative to the coordinate axes and to other pieces. These references can be features of the piece, as in a flat spot or edge, or they can be pins or clamps that are attached to the piece. Figure 5.30 shows some of the reference points used on a section of sheet metal. Some main references must be used throughout the production process; other secondary references are used only in assembly. Large assemblies, such as cars, require many references and a detailed study of the design of how each piece goes together. Each stage in the production process should maintain the reference definitions. Stations that maintain the definitions are known as geometry conforming.

The quantity, location, and strength of each weld and the weld pattern as a whole are determined by the product designer from research, design studies, and testing in the product design stage for each model. The selection of the location of tack welds concerns the number of points required for the assembly geometry and the relative position of the robot welding gun with respect to the references of the workpiece. The order that the parts are loaded on the line must allow for maximum access of the welding gun. Each stage of assembly has different access problems. Define the grouping of spot welds that can be welded by a single robot station. The time required to weld the group must take into account the position and sequence in the process and the existing tool layout. The numbers and configuration of the various welding tools necessary must be defined along with the trajectories for each and the cycle timing. If some welds are impossible to do because of the methods or the assembly, they will become apparent when the tool trajectories are designed.

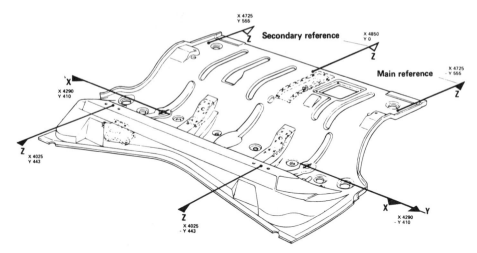

Figure 5.30 Main and secondary dimensional references for a floor panel

The overall production rate determines the design and components on the assembly line. The global line utilization factor is a function of all the elements entering into the assembly process, the workpiece supply, tool usage, and the finished product. Taking these parameters into account determines the cycle rate for all operations in the process. Handling and layout time must be considered to determine the time available for each welding station. Time constraints can necessitate parallel lines to complete large numbers of welds. All these factors go together to determine the minimum number of robots required.

The flexibility of an assembly line is defined by its degree of adaptability to the various products that are to be processed on that line. Flexibility must take into account the suitability of tooling for several different products, the capacity for adaptation, and the time needed for changeover. High degrees of flexibility may require a large investment in sophisticated machinery. A line can be initially designed to complete one type of process with allowances made for limited retooling at the end of a production run. Other lines are designed for adaptability to produce items within a range. True flexibility means being able to produce a variety of items in any order with no preliminary adjustments. The items must be similar in design and in production technique. Replacement flexibility refers to the ability of the line to compensate for the failure of one or more robots. It's a good idea to design in overcapacity or such features as rework robots to allow replacement flexibility in case of emergency maintenance.

The position of the part during assembly is very important in determining machine-loading requirements. Large assemblies should maintain their natural orientation to prestress them as they will be stressed in use. Smaller assemblies can be turned upside down or otherwise manipulated to facilitate easier access to weld points and mounting of smaller pieces. Large parts can be mounted on skids or carriages that must be handled as part of the assembly. In many cases, these carriages are used throughout the process and are important in maintaining geometric conformance. Others are used to support subassemblies during the process and cannot be removed until the process is complete.

The position and orientation of the robot with respect to the workpiece is determined by the distribution of weld points on the piece. The robot can be mounted on the floor, on a mobile or fixed base, inclined, or suspended overhead. A single gantry can span across the assembly line and support several robots. Gantry robots are especially suited to installation where floor space for the robot and equipment is limited. A typical gantry 4 to 5 meters wide can replace two floor-mounted robots that each take up 4 to 5 meters on either side of the line.

The geometric pattern of the welds in a specific application along with economic factors determine the type of robot to be used. Although you may not intend to manufacture cars, the welding of automobiles is used here to illustrate the basic types of spot welding, because autos are one of the most common (and complex) items to construct. The basic types of spot welding are the same for other applications.

In assembly processes on pieces as large as a car body panel, the references used to geometrically conform the panels get in the way of the robotic arm and welding gun. Because of the shape of a car, the references must be external to the body. These factors constrain the use of spherical robots and favor polar or Cartesian robots, because the arm often must go through a small opening to work on the inside of a cavity. Usually six robots are used for this initial assembly; two on each side of the station, one in front and one in back. It is important in these applications that the robot hand and welding gun be as compact as possible and that the three wrist axes be arranged to minimize their sweep. Figure 5.31 shows a multiple-robot assembly line. All the robots are designed to fit certain tasks and are arranged to perform the work in a specific order, so the built-up workpieces do not interfere with downstream processes.

In working on foundation chassis, such as car body platforms, all welds must be done from the outside, and the robot must be chosen for its ability to lift large heavy welding guns with long throat depths. Cartesian robots are preferred for this application. Gantry configurations allow the robot to weld symmetrical points by lateral penetrations, and a pivoting gantry robot has the maximum flexibility. When welding body side or floor frames with

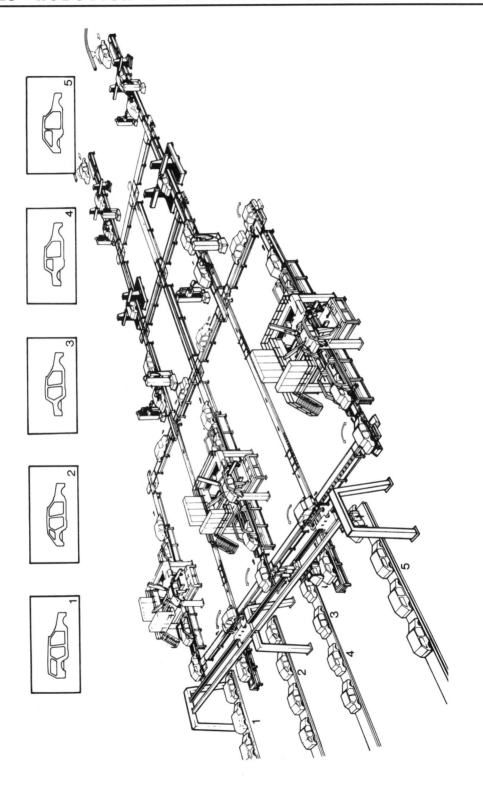

Figure 5.31 Five-model assembly line with three confrontation and tack-welding stations, each with six robots, followed by two robotic finishing lines

Arc-Welding Tasks for Robots

Arc welding is more complex than spot welding (Figure 5.32). The electrodes do not maintain a constant shape or length, and the path of the weld is crucial. Unlike spot welds, arc welds must be made while the electrodes move versus each other. For this reason, robots require highly accurate positioning devices in order to perform arc welding (Figure 5.33). These positioners can be as complex as the robot, and the positioner must be able to synchronize its activity with the robot, usually through a shared central control computer.

Figure 5.32 Complete set or arc-welding robot station and the unit to program its control

large openings, access can be from the side of the body as well as from the center of the opening. Cartesian robots are by far the best choice for this particular application, because the weld patterns are often straight and parallel to assembly-line motion. For vertical planes and lines, such as those in a car's side panel when the car is

(a) *ESAB Orbit 500*

Workpiece:
    Max weight   1100 lb (500 kg)
    Max diameter   57 in. (1460 mm)
Indexing: 90°   360°
    Rotation   2.7 sec   7.2 sec
    Tilting   3.3 sec   8.9 sec

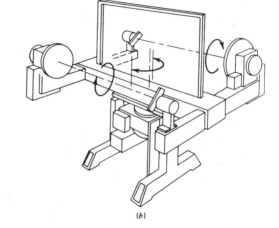

(a)

(b) *ESAB Orbit 160 R*

Workpiece:
    Max weight   352 lb (160 kg)
    Max diameter 45 in. (1150 mm)
Indexing: 90°   360°
    Rotation 2.3 sec   6.8 sec
Station interchange (180°): 4.5 sec

(b)

(c) *ESAB Orbit 160 RR*

Workpiece:
    Max weight   352 lb (160 kg)
    Max diameter   45 in. (1150 mm)
Indexing: 90°   360°
    Rotation 2.3 sec 6.8 sec
    Tilting   2.7 sec 7.2 sec

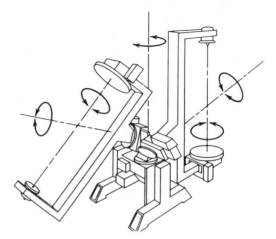

Figure 5.33 Five types of part positioners for arc-welding robots: (a) ESAB Orbit 500, (b) ESAB Orbit 160R, (c) ESAB Orbit RR.

(d)   ESAB Orbit MHS 150

Workpiece:
   Max weight   330 lb (150 kg)
   Max diameter   45 in. (1150 mm)
Indexing: 90°   360°
   Rotation 6.0 sec 12 sec
   Tilting   4.0 sec   —
Station interchange (180°): 7.0 séc

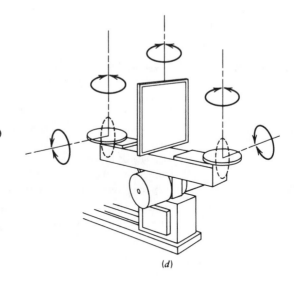

(d)

(e)   ESAB Orbit MHS 500

Workpiece:
   Max weight   1100 lb (500 kg)
   Max diameter Any
Indexing: 90°   360°
   Rotation 11 sec 22 sec
   Tilting   5.0 sec   —

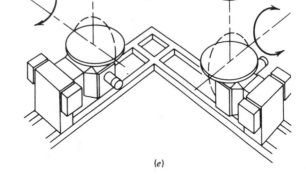

(e)

Figure 5.33 (continued) (d) ESAB Orbit MHS 1503, (e) ESAB Orbit MHS 500

in its natural position, gantry robots are difficult to use. Most other configurations are satisfactory, especially those with linear penetration to reach points inside the body. All types of robots can be used in final assembly lines and are selected primarily on the basis of access and flexibility.

CHAPTER
6

Assembly

After all the pieces have been designed and fabricated, it is time to put them together. Assembly is the next sector in the manufacturing cycle. Even at automated fabrication centers like the John Deere plant, robots have not replaced the assembly-line workers. Assembly tasks are extremely challenging to robots. This chapter explores the nature of general assembly tasks and the abilities and limits of robots in replacing human beings. The assembly center shown in Figure 6.1 is an example of a sophisticated system that includes two robots with vision systems.

Computer-controlled pick-and-place assembly robots are the simplest kind of robot. They often have limited degrees of freedom, their movements are simple, and they can be quickly reprogrammed to work in different assembly configurations. The Japanese have made this type of robot their main investment: 70 percent of their 50,000 robots are of the pick-and-place type. Only 40 percent of America's 15,000 robots are pick-and-place robots.

Robots with six ranges of motion can mimic the movements of a human. These more sophisticated robots are generally used

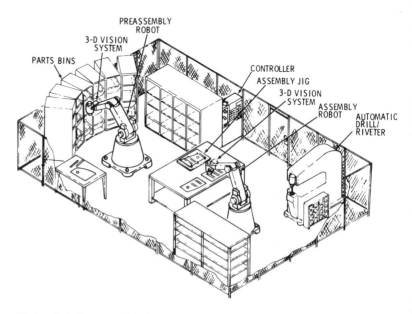

Figure 6.1 Conceptual design of advanced robotic assembly center

for welding and spray-painting applications, but they have some of the same limitations as a human when considering joint movement capabilities. They move no faster than human arms can, they are only as precise as the computer control program and their environment allows them to be, and they are limited in the loads that they can carry without stress failure. If an error puts the workpiece in the wrong place, the robot continues on mindlessly, never knowing the difference. Robots are, in general, more expensive than people. Depending on the type and complexity, robots cost $50,000 and up, with an additional $100,000 to install the robot and make it operational. At best, the robot will perform the same work as one or two humans. No matter how efficient robots are, on a cost basis, they simply cannot compete against humans unless the labor rates are high. The real advantage is the robot's ability to continue working for long periods.

But, even if the costs were the same, can assembly tasks be defined well enough to be done by robots? Many tasks that are very easy for humans to perform, such as threading a nut on a bolt, are extremely difficult operations for a robot. When a human performs this simple task, the fingers use a complex sensing network to detect forces accurately. If the nut is not threaded properly, the human can detect this condition and compensate for it until the correct action is performed. Humans do this sort of thing by second nature, and you don't even think about the tactile-sensing ability that enables you to accomplish such simple tasks effortlessly. Robots cannot claim such abilities; in fact, if something goes wrong, for instance, if the nut is threaded wrong, the robot continues to thread the nut, forcing it onto the bolt. Robot designs would have to include sensing to deliver a tremendous amount of information in order to instruct the robot's mechanical fingers in the correct procedure. Even so, duplicating everything that human beings can do is not going to address manufacturing issues, because we already have humans that can do these tasks. In many specialized and awkward welding tasks on the automobile assembly lines in Detroit, human welders still work alongside robots. The numerous cables and wiring connections that are necessary for automobile and truck electrical systems are another case where there is no substitute for the human mind and hands. Figures 6.2 through 6.4 illustrate typical assembly tasks.

Human behavior and physical attributes are not what is really needed in a manufacturing environment. The concept of a complex universally adaptable robot comes directly from artificial intelligence laboratories where the goal has been to copy human behavior. It is important to make a distinction between sophisticated laboratory robots and the robots required

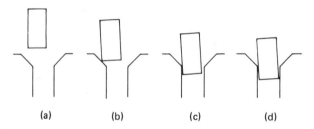

Figure 6.2 An engineering approach to the four stages of parts mating

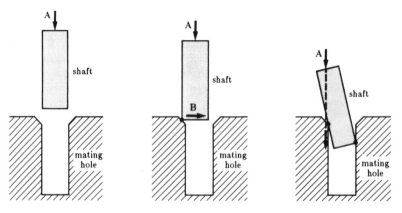

Figure 6.3 The free-body diagram showing the forces involved

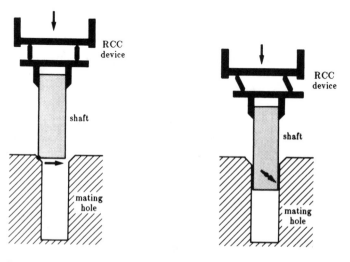

Figure 6.4 A remote center compliant (RCC) device accounts for the forces to automatically align the piece with the hole

to perform specific functions with the "bottom line" as the chief factor in judging performance. The research in universities and laboratories develops computer automation and robotics, supplying industry with the leading edge of technology. But until incredible breakthroughs in robotics costs and abilities occur, humans are likely to remain more cost-effective than any complex, expensive robot.

One way to achieve more automation with less sophistication is to design products for easier assembly. Many companies are putting a lot of effort into the designing of new parts as well as redesigning old parts, so that simple pick-and-place robots can be used cost-effectively in the manufacture of those parts. In Lexington, Kentucky, IBM has set up a $350 million automation plant. They have redesigned their new electronic typewriters so that they can be assembled using robots and computer automation. The parts to be assembled are presorted, so the robots need only pick the part and place it in the proper position.

Designing for robot assembly is a radical departure from the traditional method of designing for efficient function and assembly by humans. After a new product is designed or an older product is redesigned, the addition of the computer's flexibility enables manufacturers to accommodate design changes quickly by modifying the computer software controlling the robot's functions. This capability brings automated flexibility into assembly, which, in turn, enhances the productivity of the installation.

Certain jobs, such as the assembly of typewriter circuit boards, can be performed by computer-controlled robots with speed and accuracy that no human can match. This brings to IBM the advantages of better quality, reliability, productivity, and lower labor costs. IBM's Lexington factory is at the vanguard of assembly automation in America, but, at present, there are very few such installations. Such a plant supplies evidence that redesigning the product for assembly by simple cost-effective robots is a productive way to go. This is not to indicate, however, that all products can be redesigned; many products do not lend themselves to automation at all. Many experts say that most automobile assembly tasks might best be left unautomated and continue to be done by humans. Advocates of all-out automation disagree. They argue that there is a need for complex robots to replace people in even the most complicated assembly jobs.

Humans have a big advantage in that they can alter their performance to account for any changes in the environment. For example, if parts do not fit properly, the human can change the way that they try to put them together. Even the most sophisticated robots cannot approach this logical adaptive flexibility.

The future will show a wide spectrum of robotics applications in America's assembly operations. Some jobs will be done by unsophisticated pick-and-place robots, while the difficult and complex assembly tasks are going to be done by more sophisticated robots. Robots will require one very important feature in order to replace humans in assembly work—eyes. Vision systems would enable automated machines to be able to select the pieces that are needed for the assembly tasks.

Assembly is simple in definition; it means putting things together to form a unit. These tasks can usually be done by a human with a minumum of hand tools. The trouble with automating such tasks lies with outfitting machines with the logic and dexterity that a human uses as second nature. Humans can look at a group of pieces and infer the assembly process, successfully completing even difficult assembly operations with no instruction. An implicit instruction such as "should fit smoothly" is enough to virtually guarantee successful assembly. At present, machines do not come close to the logic of a human in these situations and cannot understand implicit instructions. Every step of a process must be explicitly laid out for a robot. How will the robot know that the parts don't fit? What is the robot going to do if the parts do not fit smoothly?

The basic issues in the design of assembly systems are economy, technical accuracy, and quality. Flexibility in assembly is required to adjust to varied models and changes in manufacturing processes. People are by far the most dexterous and flexible assembly systems, but their performance varies, and the cost is high. Fixed automation provides the greatest economic benefit for large production runs, but it can't provide the flexibility required to produce small batches economically.

A programmable assembly system, such as robotics, can offer an alternative somewhere between the extremes of manual assembly and fixed automation. Robots incorporate some of the dexterity and flexibility of people and offer the economy and efficiency of dedicated fixed automation. However, assembly presents perhaps the greatest challenge to the dexterity of robots, and an extreme amount of preparation must precede robots in the assembly process. Robots must be ordered to perform every detailed move, and many jobs have not been investigated and documented to the extent required to program robots to accomplish the task. The automated programmable assembly system consists of a number of assembly stations interconnected by parts transfer mechanisms and tied together with a central controller. Each station consists of dedicated equipment served by people, robots, programmable feeders, magazines, and fixtures all synchronized by the central computer.

Figure 6.5 shows six layouts for different types of auto-mated assembly arrangements. Figure 6.6 shows a more compli-cated complete system in actual use by Hitachi of Japan in the production of video tape drives. View (a) is of the complete installation, showing all the stations in the assembly line and the inductively guided self-driven delivery vehicle. View (b) shows a close-up of one of the assembly robot stations. The magazine supply fixtures are shown in the process of being exchanged by the self-driven vehicle. Figure 6.7 shows a programmable in-line assembly system with magazines, feeders, presses, and robots.

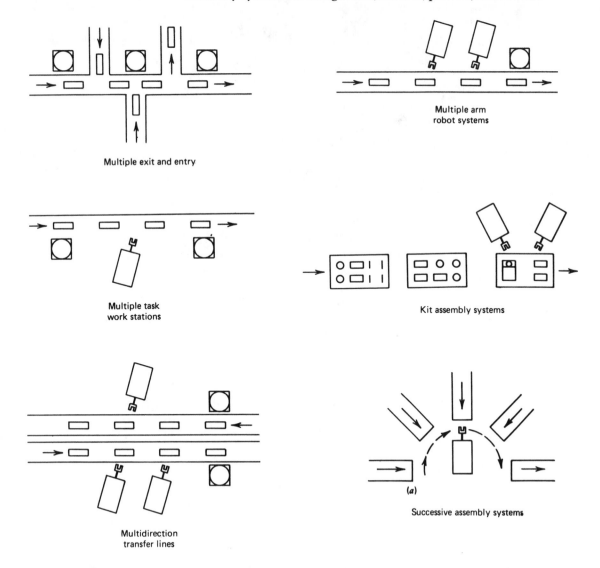

Multiple exit and entry

Multiple arm
robot systems

Multiple task
work stations

Kit assembly systems

Multidirection
transfer lines

Successive assembly systems

Figure 6.5 Typical configuration of robotic workplaces including flow

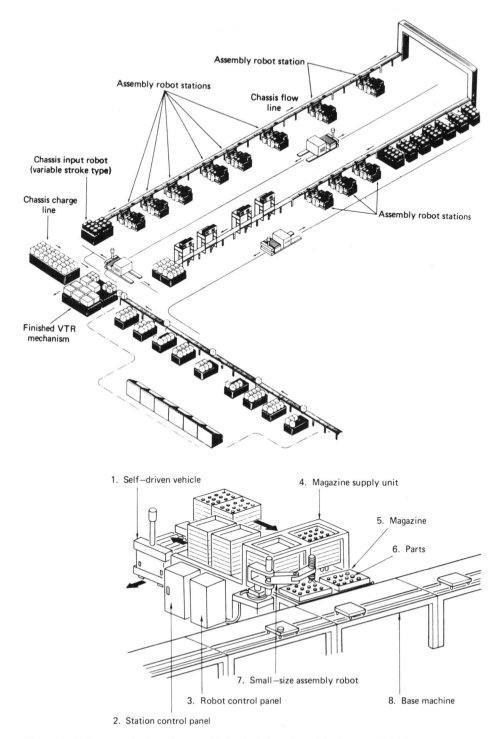

Figure 6.6 Diagram of automatic assembly line including view of single assembly station

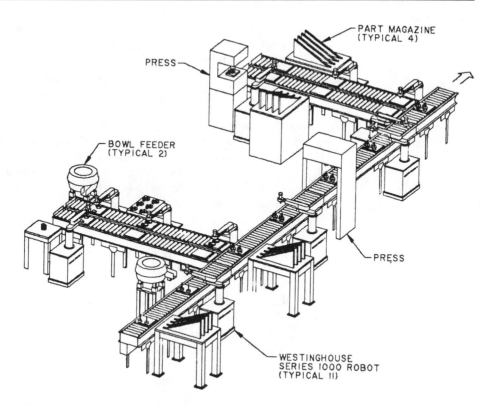

Figure 6.7 Programmable in-line assembly system

Planning the Assembly Cell

Implementation of robots in assembly operations should follow a comprehensive plan that identifies every detail of the proposed operation. Whether the program is rigorously analytical or trial and error, each detail will surface as a problem to be solved before production can begin. The typical design process includes the following steps:

1. Project initiation and approval. Initiation and approval depend on the economic benefit to be gained or the removal of workers from dangerous working conditions.
2. Candidate product survey. Dangerous jobs can be readily identified from accident reports, but the selection of the application of cost-saving automation is more complex. It is necessary to consider the yearly volume of product, the number of different styles within a product group, and the assembly time per unit. Related factors are the volume of each style, the number of process changeovers per unit time, the complexity of each changeover, the number of parts in the assembly, and the major part dimensions (weight, width, length, etc.).

3. Preliminary product screening. The same criteria as in the initial survey are applied more rigorously, eliminating some candidates. A simple economic justification uses the annual volume and number of product styles to determine costs for dedicated, programmable, and manual assembly methods.

4. Prioritization of candidate products. Candidates are prioritized according to economic benefit, technical difficulty, and environmental, social, and political impact.

5. Project team selection. Regardless of the depth of the initial surveys, the information gained will not be sufficient to actually implement the design. This is where hard engineering takes over. Robotics expertise is hard to come by even in the largest facility, so it may be best to call in the experts. Assign a team to work directly with the vendor to install and test the system. This develops in-house expertise painlessly and furnishes a direct communication line between you and the supplier.

6. Data collection and documentation. Collect information on the product. The monthly and yearly production volumes and number of styles must be verified and monitored. Consolidate engineering drawings for each style of the completely assembled product as well as all parts in the assembly. Include the engineering drawings and specifications on all the equipment, tooling, and fixturing used to assemble the product. Obtain instructional material for the assembly sequence and methods. Reports on cycle times, average batch size, defect types, and quality need to be compiled, along with all testing and inspection data gathered from before and after the assembly process.

7. Conceptualization of configuration. The final concept includes detailed specification of the final layout and dimensions of the system components. The sequence of all assembly and inspection tasks must be determined. The work envelope must be designed, laying out the work stage, transfer mechanisms, assembly fixtures, feeder, grippers, special tooling, and testing devices. The part feeders, dedicated equipment, and storage buffers must be placed at the edge of the work envelope.

8. Economic evaluation. After the major parameters have been developed, an extensive economic evaluation can be performed. The areas impacted are the total quality, floor space in the shop, net yield, warranty costs, and in-process inventory.

9. Detailed design. The detailed design of the system must include studies of the assembly process and the kinetic analysis of each movement and operation together with the control systems needed to synchronize all activities. The components of a programmable assembly system are: the workstation, the robots and end effectors, the feeders and transfer mechanisms, assembly fixtures, and sensors.

Assembly Equipment

The Workstation

The function of the workstation is to provide a stable, well-defined place for the assembly and inspection tasks. The components are the station substructure (flooring, concrete, tables, etc.), tooling platform, and locating devices. Delivery of utilities, such as electrical signal and power lines, gas lines, and hydraulic and vacuum lines, is often a problem. Carefully design the available space in the workstation. Figure 6.8 shows two robots at work in a typical assembly station.

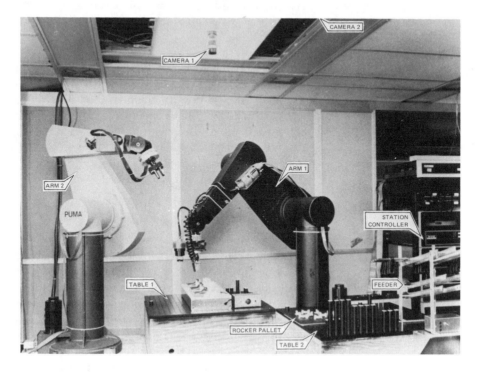

Figure 6.8 Typical assembly station

The Robots

The robot must have a large enough work envelope to accommodate all the machines, conveyors, and so on required to perform the assembly task. Consider the robot's repeatability, accuracy, payload, speed, degrees of freedom, and control system. Design the foundation of the robot to occupy minimum space, but make sure it is able to support the robot and absorb vibrations from the natural frequencies developed by the robot motions.

The End Effectors

The operational efficiency of the end effector plays a major role in the efficiency of the entire system. Design of the end effector is much like designing tooling for other types of machines, taking into account the workpiece, the task, and the environment. In a programmable assembly system, one of the best ways to reduce cycle time is to have the same end effector handle all parts in the station. The advantages of the multipurpose gripper are that tool changing time is eliminated and parts transfer time is reduced. The disadvantages are that they are more difficult to design and more complex in operation.

The Parts Feeders

The feeder must be able to provide parts, separate them, orient them, and present them to the assembly mechanism at a known location. These are usually complex machines themselves. The rate of delivery and the space required are crucial to the system as a whole. Figure 6.9 shows a parts feeder that can take parts loaded in random fashion and present each part to a robot right side up and in a known location.

The Transfer Mechanisms

Product in transit is work-in-process inventory and does not increase in value by the move. In fact, it costs money to transport items, and they could be damaged or disoriented. Whenever possible, transfer mechanisms should include a storage buffering capability. Such mechanisms should be asynchronous with respect to the assembly process in order to absorb work slowdowns or machine malfunctions. Figure 6.10 shows a robot used as the transfer mechanism within an assembly cell. Robots are very good transfer mechanisms, because they retain any required orientation of the parts and are programmable.

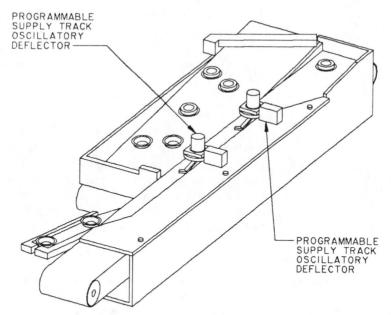

Figure 6.9 Programmable belt feeder

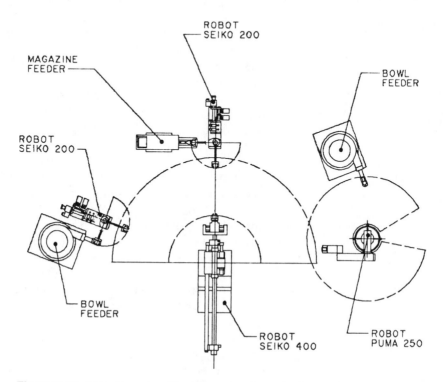

Figure 6.10 Assembly center with robot as a transfer mechanism

The Assembly Fixtures

The role of the fixture is to support the work in process. The fixture should be adaptable or interchangeable with other fixtures to enable work on different workpieces. Many times the assembly fixture is another robot.

The Sensors

Even simple pick-and-place operations require complex motion, and the ability to sense the position of parts as well as gripper attitude (open or closed). Sensors are needed to assure that the assembly sequence takes place as planned and that deviations are detected in time to do something about them. Servo-control mechanisms have internal means to monitor location/orientation and can take the place of some sensing equipment.

General Concepts in Assembly

A considerable amount of data on average cost and performance can be gained from the large installed base of robots and special machines. Compare the per-unit costs of production for various production volumes and for different assembly techniques. Parametric models can supply crucial information on the average cost of buying, installing, and operating the robots, plus their productivity or operating speed. Many aspects of the basic system can be developed directly from computer simulations and systems synthesis tools.

Figure 6.11 shows a layout for assembly of automobile shock absorbers. Each parts feeder is carefully placed in the cell to allow the robot (not shown) to reach each one. Figure 6.12 shows the same shock-absorber assembly cell, identifying the crucial points above each of the feeders and assembly posts.

Analyze the process methods with emphasis on difficult operations and awkward sequences. Keep an eye out for operations that look easy when done by hand but are difficult to program. Be prepared to replan any subassembly or upstream process that causes trouble downline. Many operations and subassemblies have never been scrutinized before, because they were judged to be either so simple as to not require thought or so fundamental that redesign was not economical. Take a close look at all subassemblies and basic chassis. A few changes in these input items can be inexpensive and make automation easier.

Before any robots, sensors, computers, or other equipment is selected, there are important issues to be resolved. The layout of machines, conveyors, fixtures, and feeders, as well as the geometry of each workcell, must be designed and compared to

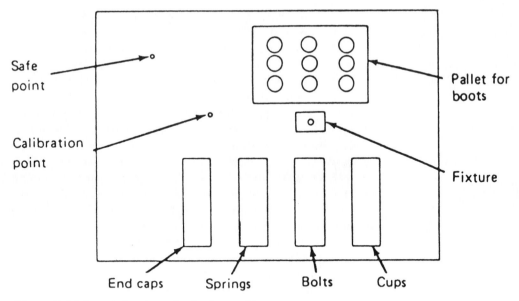

Figure 6.11 Layout of work area for shock assembly

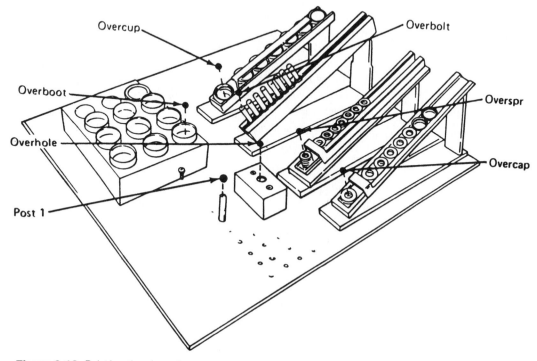

Figure 6.12 Point locations in work area

the alternatives. The extent, or travel, of each move as well as direction must be taken into account. The tolerances required on all motions, speeds, forces, vibrations, and noise must also be known. Don't mistake these steps for unnecessary digressions. All moves, however small, should be part of a design that eliminates unneeded motion. This makes each remaining move critical. A large degree of vagueness is permissible in a manual operation, but not in the specification of motion for a robot. Qualities such as smoothness cannot be judged by a machine. All motions and operations must be explicitly stated.

Compare all likely scenarios for each operation, defining all requirements for computers, robots, and tools, both economically and technically. Experiment with the robots to determine their exact operation. Performance standards, such as maximum effector speed and repeatability, are useful when comparing robots, but may not be directly applicable to the actual task. In addition, the robot may be exposed to heat or other factors that affect performance. Stop-to-stop time or settling time may be more important than actual peak velocity. Absolute accuracy in a one-time motion is different from repeatability. Most robots can perform the same action over and over, but this action requires only that the internal coordinate system of the robot be consistent from move to move. Absolute accuracy requires that the internal coordinate system exactly match the real world. This requires an internal calibration on the part of the robot and external reference points, such as consistent part placement. Some operations require adjustable resolution and the ability to make small moves to compensate for errors.

Planning the alternatives and selecting the best solution to a given assembly task is highly specific, and there are no substitutes for hard engineering and economic analysis. If different layouts are possible, each should be investigated. The luxuries of expansive floor space and unlimited budgets can be enough impetus to design sweeping changes, but, in general, optimization of these resources must be considered with projected and realistic production volumes. Figures 6.13 and 6.14 show two vastly different layouts for the production of the same item. The two layouts compare production for different robots and feeders as well as different configurations.

Extend the process modeling to cover the actual performance and evaluation of the robot. Although kinematic and dynamic models of idealized robots differ greatly from the real-world performance of the robot, the differences between the real and ideal situations can be computed, yielding the performance analysis needed to calibrate the robot and anticipate inaccuracies. Increased automatic control capacity, together with

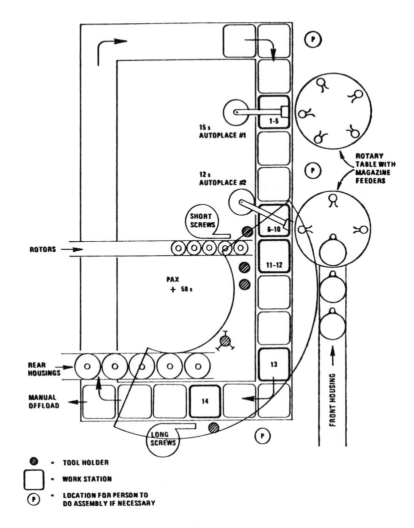

Figure 6.13 Solution A to the production requirements can produce 124,000 assemblies per shift-year

better sensing, will enable the robot to monitor its performance and calibrate itself to adjust to inaccuracies.

The best way to deal with errors is to use a sensing system that enables the robot to make changes in its motion to get back on track. Choosing the sensing system is crucial to technical and economic success. If mechanical stops, fixtures, jigs, or guides can be used, it makes little sense to invest in expensive vision systems or tactility sensors. Compliant wrists and fixtures can absorb small errors where the robot need not follow an exact path. Such systems can improve reliability, reduce costs, and lower system complexity. Very small contact-no-contact sensors

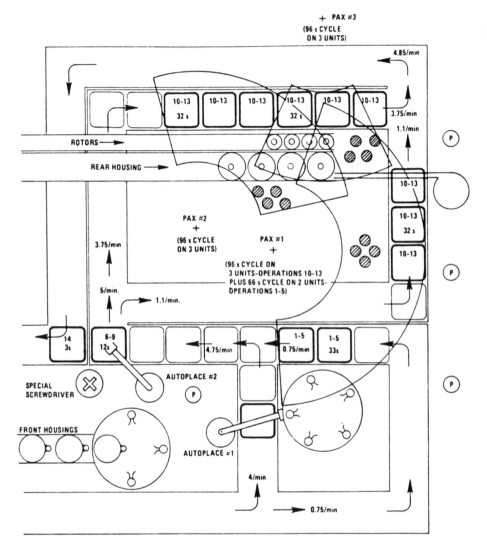

Figure 6.14 Solution B to the production requirements can produce 544,000 assemblies per shift-year

are cheap and easy to install. Many of these devices can be used in place of a large sophisticated and expensive vision system. Force /torque sensors in the robot wrist can detect overload conditions and keep the robot from damaging itself against an obstacle or from damaging the workpiece due to improper fit.

CHAPTER
7

Inspection

The ability of a computer to recognize objects is limited, and the science that deals with feature recognition in a jumbled picture, such as that of a random pile of assembly parts, still has a long way to go. But computer recognition of simple distinct shapes is already being used in the next sector of manufacturing—inspection. This chapter briefly covers some of the basic concepts used in the vision systems that are being applied in the factory today, enabling computers to gain information about their environment. Increases in the capacity of vision systems and accompanying software to adapt to different situations are primary factors in the expansion of robots in manufacturing.

Vision controls robots that sort bolts on conveyors and guides a robotic welder in a toxic environment so thick with smoke that a human could not see, much less breathe. Modern advances in image interpretation and array processing software techniques have pushed vision systems out of the laboratory and into the factory. Since 1978, vision system vendors have increased in number from fewer than 10 to more than 100. The industry accounted for more than $85 million, and industry watchers project a market of $1.8 billion by 1994. Figure 7.1 shows a robot verifying the application of sealant to a car body frame.

The IBM facility in Lexington, Kentucky uses computer vision systems to inspect new typewriters that have been assembled using robotic automation. This is a simple computer vision system: the required visual recognition is limited to the letters of the alphabet and other typewriter symbols. The test station prints all the characters in specific combinations and compares the pictures it takes of the typed characters to established standard pictures of what the characters should look like. It is designed to work simply, as a part of the whole approach of redesigning the product for programmable automation.

The concept of product quality can be defined as the totality of features and characteristics of the product that bear on its ability to satisfy a given function. A balance must be found between the cost of achieving a specified quality level and the cost benefit of achieving that level. Inspection is defined as the process of identifying pieces that do not fall within the specified quality range. Inspection prevents nonconforming pieces or material from proceeding further in the process and is used to gather information on specific characteristics of parts or materials to determine changes in the manufacturing process. Inspection usually means measuring geometric dimensions, surface finish, position accuracy, and assembly integrity. All these tests can be performed by humans just by looking at the workpiece, making vision a prime requisite for machine inspection systems.

Simple devices have been used for years to determine the presence and absence of certain parts. More sophisticated systems emulate human vision, using stereo or matrix array cameras and structured light to match a three-dimensional representation with an established standard of perfection. The simplest are one-dimensional line scan cameras, but two-dimensional pictures are commonly used as well. The science of analyzing these pictures and putting them in a form that computers can handle is a part of the study of artificial intelligence (AI). Artificial Intelligence is a complex and young computer-programming field in which attempts are made to give machines the ability to "think." You don't need an in-depth study of artificial intelligence to understand machine vision, however.

The basics of machine vision can be understood by analyzing a human's thinking processes when presented with a complex

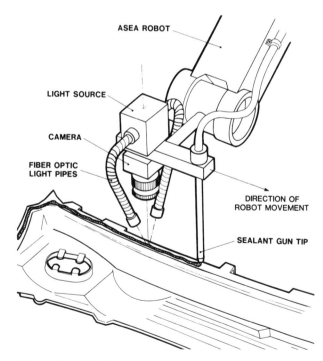

Figure 7.1 Robotic inspection of sealant application

scene. People always look for things they recognize in a scene. They look at the clouds in the sky and find one that looks like Abraham Lincoln, or they find their mother's face in an ink blot. When you look into a box of parts, your mind compares the picture to every possible orientation for the part, finding ones that are upside down, backwards, or partially hidden and making instantaneous recognitions. The computer cannot make these recognitions as fast as a person, so the real secret of implementing computers in inspection tasks is to limit the number of orientations that the computer will see. If the workpiece is placed in a known location and orientation, the computer only has to make one comparison and one recognition. Figure 7.2 shows a sophisticated robotic workstation guided by a vision system powerful enough to process the incredibly complex images in the scene.

The IBM typewriter example is simple, because each letter is supposed to appear in a certain location and be of a specified shape. All the computer has to do is determine whether the character did indeed appear as was expected. Any deviation from the expectation is recorded. This is a simple case of go–no go, where physical presence or nonpresence is detected visually, rather than mechanically. This method is used in sophisticated ways, such as spreading light around the picture in special ways and placing cameras in special positions, or by using special cameras, such as ultraviolet or infrared cameras.

Vision Systems

The two main components of all vision systems are the camera and the computer. The camera, just like the human eye, is sensitive to the light reflected off the objects in a scene. It records

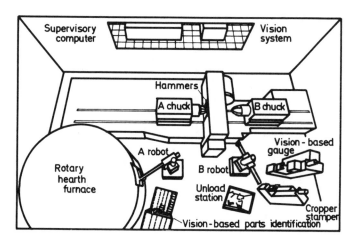

Figure 7.2 Robotic workcell controlled by a vision system

Inspection in Action

Figure 7.3 shows two images that the computer compares to determine imperfections. Pattern g is the good pattern and pattern f is the faulty pattern that was photographed. Figure 7.4 shows the process of extracting all the information derived from the comparison of the patterns in Figure 7.3. Figure 7.5 shows (a) a target pattern, (b) the ideal, or comparison pattern, and (c) the result of comparing the two with the differences between them. Figure 7.6 shows the schematic diagram for the system the check exemplified in Figure 7.5. Note that the stage holds two boards. This arrangement can be used to check one board against a known board, as shown, or it can be rotated to

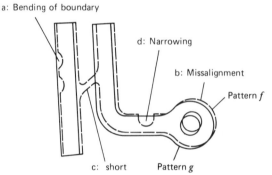

a: Bending of boundary

d: Narrowing

b: Missalignment

Pattern f

c: short Pattern g

Figure 7.3 Overlapped good and faulty patterns

	Pattern f and its processed image	Pattern g and its processed image	Result of comparison of F and G	Configuration of feature extraction operators
(1) Detected patterns f, g				
(2) Extracted boundary lines F_K, G_K in the Y direction	F_K	G_K		
(3) Extracted fine line patterns F_B, G_B in the X direction	F_B	G_B		

Figure 7.4 Process of feature extraction

(continued)

change the scene for each of the two cameras. Figure 7.7 shows the actual installation of an automatic inspection system for printed circuit boards.

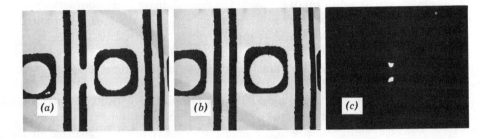

Figure 7.5 Pattern comparison

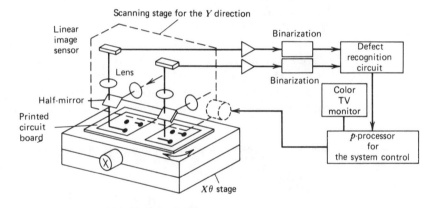

Figure 7.6 Schematic diagram of comparison system

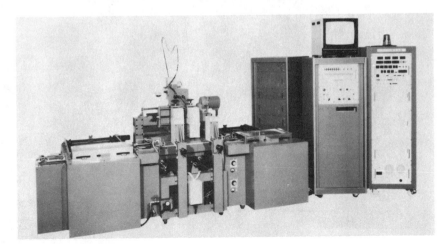

Figure 7.7 Complete automatic inspection system

the image as variations in the light received from various angles and generates a pattern of voltage values. The computer interprets the pattern as a matrix of picture elements, or pixels. The resolution of the system is determined by the number of pixels in the matrix array. Depending on application, the system resolution can vary from 10×32 pixels in the square array to 2048×2048 pixels. Figure 7.8 shows an inspection system consisting of the stage, two light sources, the camera, and the computer.

Three computational methods are used to interpret the information from the camera: image buffering, edge detection, and windowing. Image buffering involves digitizing an entire image and storing the digital information where a computer program can analyze the image for features such as perimeter, area, centroid location, and orientation. Edge detection systems do not require acquisition of an entire image; they use software to compute the features of a scene from the locations of transition from black to white and white to black in the image. Through a process of connecting these locations to form "blobs," the features of the scene are determined. Windowing further reduces the amount of information required to analyze a scene, by selecting only certain areas to be scanned. These areas, or windows,

Figure 7.8 A simple vision system

Stereo and Nonstereo Imaging

Figure 7.9 shows a grey-scale picture of two stools. The one on the right is slightly farther from the viewer. It isn't easy for a human to get used to seeing things this way, but it is possible to discern the basic shapes.

Figure 7.10 shows a stereo-matched pair of pictures of the two stools. This picture would be enough for a person to know where the stools were, be able to negotiate between them, and even sit down on one of them.

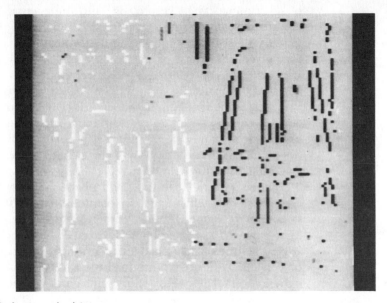

Figure 7.9 A grey-scale picture

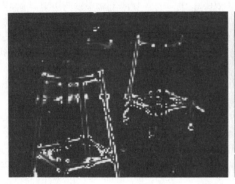

Figure 7.10 Stereo grey-scale pictures

can include a weld point or a hole that is the actual target for inspection, ignoring other features that do not require inspection.

Each of the pixels in an array can be assigned a value. A binary picture contains pixels that have one of two possible values, on or off, that is, white or black. All light values in a scene are reduced to black and white pixels that have a digital value of 1 (on) or 0 (off). Other systems are set up to give an intensity value to each pixel, in order to discriminate up to 256 shades of grey. Such a grey-scale picture has a discrimination greater than that of the human eye. Most applications will not require a high degree of discrimination or resolution, and more complex pictures require more processing power in the computer. A good compromise uses just the resolution and discrimination that is required for a given task. These parameters are usually determined by the software for the computer, and the camera can usually be adjusted automatically.

The amount of light in the scene is critical to the camera, of course, but the structure and direction of the light is equally important. A grey-scale picture can be used to tell if a painting operation has been done only if the camera knows the amount and intensity of the light reflected off the painted surface. Figure 7.11 shows a robotic painting and inspection line with two cameras checking each side of the painted box. Light can also be structured to illuminate only certain features of a scene. When light is limited to a line of light focused on an object at a certain angle, the line of light changes according to the contour of the object. Three different ways to check surface contour using the light section technique are shown in Figure 7.12.

Three-dimensional pictures of objects can be derived from one camera by using structured light. In Figure 7.13, structured light is used to reveal the features of a cylinder on a flat background. Depending on the angle between the light source and the camera, the single light spot in view (a) jumps when it leaves the flat background and hits the cylinder or disappears from view when a boundary is reached. These actions in a time sequence of photos determine the boundaries and contour of the cylinder. In view (b), a bar of light is aimed at the cylinder from a specific angle. Perpendicular scan lines record the position of the bar. As the light bar is moved from left to right across the scene, the scan lines pick up different illuminated points, revealing the contour of the cylinder. In view (c), a grid of horizontal and vertical light bars shows the contour of the cylinder in a single picture. Of course, there is quite a bit of processing necessary to understand

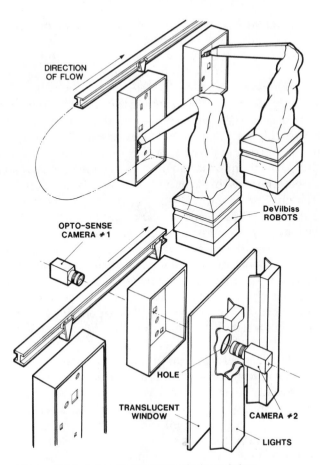

DIRECTION
OF FLOW

DeVilbiss
ROBOTS

OPTO-SENSE
CAMERA #1

HOLE

TRANSLUCENT
WINDOW

CAMERA #2

LIGHTS

Figure 7.11 Automatic inspection of robot painting

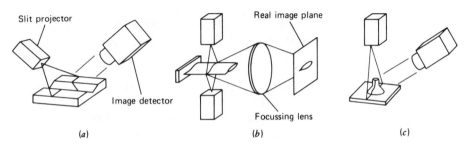

Slit projector

Image detector

Real image plane

Focussing lens

(a)   *(b)*   *(c)*

Figure 7.12 Three light section methods

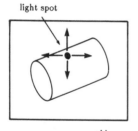

light spot

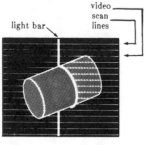

video
scan
lines

light bar

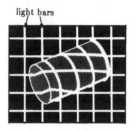

light bars

Figure 7.13
Structured light patterns for
stereo range-mapping
techniques

the contents of the scene. A human can look at these pictures and understand the scene, but the computer must have software that tells it how to interpret every arc and line and how it deviates from the picture of the flat background.

Lasers can be used to detect flaws in what should be a perfectly flat surface by scanning the surface and having a computer measure the degree of scatter from the irregularities in the surface. Lighting from behind can reveal silhouettes, and strobe lights can freeze movements to reveal details in dynamic structures, but this information must still be interpreted. One method for determining the structure of shapes is illustrated in Figures 7.14 and 7.15. Figure 7.14 shows a connecting rod and a hexagonal nut. Each shape is shown as one or a series of blobs separated by a contrasting color. Mathematical processes of connectivity separate blobs that are linked together from noise pixels resulting from dust on the scene or from electronic noise in the detection circuitry. The connecting rod is digitized according to the intensity of darkness, and each pixel is assigned a numerical value, as shown in Figure 7.15. The resulting digital skeleton of the object is reduced to only those bits of information needed to identify the object against its background.

After the information in a scene is obtained, the computer program determines the resulting action. Sometimes the item is merely counted or diverted onto a conveyor and sent to a storage area. A vision system functions primarily as a sensor, taking no action itself but relaying information to the plant's central computer. A nearby robot is sometimes used to load items onto an inspection table and put the inspected item in one of two places, depending on the result of the test.

Applications of Vision Systems

As of 1984, there were only a few hundred machine vision installations, so the applications have barely been touched, but there are some solid trends developing. About 47 percent of vision systems are used in the verification, gauging, and inspection of work for flaws. Computer vision is ideal for application in these areas. The model of perfection for the workpiece is developed, and the computer makes comparisons between the ideal and the test piece very quickly, depending on image content and complexity. Computers can work steadily, with no differences from test to test, achieving close to 100 percent accuracy of fault detection. Humans, even when they are not bored to death by looking at the same thing over and over, rarely achieve a detection rate of 90 percent. Inspection tasks include such activities as detecting the presence of a piece or an assembly, checking the contents of a container, and verification of the contours of a

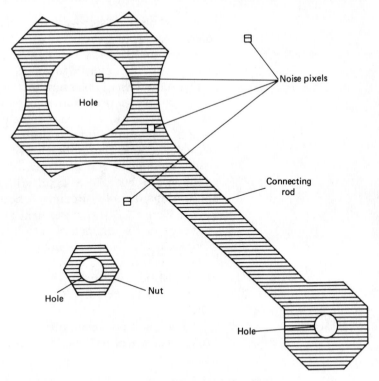

Figure 7.14 A binary picture of a connecting rod and hex nut

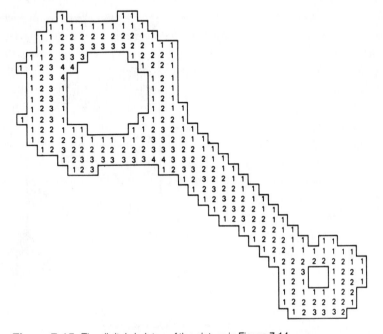

Figure 7.15 The digital skeleton of the picture in Figure 7.14

machined piece. Vision systems can measure to extremely high tolerances, making them viable for the gauging of small items, such as rotor blades in a jet engine and the material etches in a silicon semiconductor. Figure 7.16 shows a layout for a robotic testing station using a test nest that incorporates a vision system.

One of the main forces behind the development of the vision systems industry has been the robot industry. Development of machine vision was primarily viewed as a necessity for advanced robots. The application of sophisticated vision systems to the guiding of robots at manufacturing tasks accounts for about 16 percent of the applications of vision systems. Such applications include directing robots to pick items from conveyors. Figure 7.17 shows a typical layout for a robot that serves a conveyor that is scanned by a vision system. The robot can be directed by the vision system to remove parts from the conveyor if they meet or pass a certain criteria. If the vision system can distinguish between many parts, the robot can be programmed to sort out whole parts families, putting each type on a different pallet.

There are also examples of vision used in more sophisticated tasks where the information gained is used to track pro-

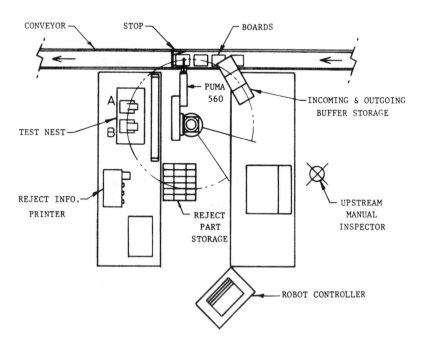

Figure 7.16 A robotic test station

cesses such as welding. The camera in this application watches the progress of the weld and guides the robot along the seam of the weld. Computer vision systems do not necessarily rely on the same visual wavelengths of light that humans can see. Cameras can detect ultraviolet or infrared light that cannot be seen by a human and look into an electric arc that would blind a human worker. There are also applications that do not use light at all; rather they use radio waves, x-rays, and acoustic signals that require sophisticated reception equipment. These systems are capable of examining the internal features of structures that are solid, such as an engine casting or a geological sample. Several years ago a priceless painting was identified by sonic scans of a supposedly worthless canvas, without destroying the overlying painting. These systems use computers to analyze the scattering of the radiation medium (sound, x-ray, etc.) and develop pictures, or maps, of subsurface variations in material density. The maps are compiled to make a picture of the features that cannot be seen from the outside. As these many applications are explored, the costs of the systems will drop, spurring even more applications.

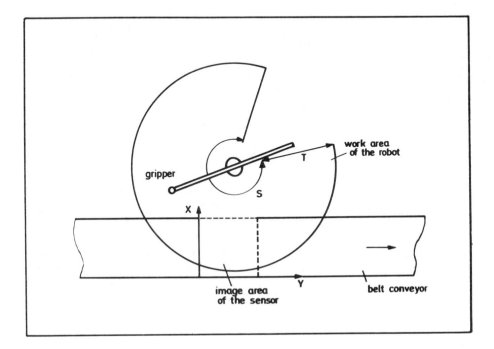

Figure 7.17 A typical layout of a robot vision system serving a moving conveyor belt

Materials Handling

How far can sophisticated technology take the factory toward automation today? The technology is available to provide great productivity and quality gains, so availability is not the problem. The main obstacle to increased factory automation is the unwillingness of management to risk large investments by taking a chance and doing things in a new way.

The $30 million that John Deere has spent on an automated materials-handling system represents such a risk. Materials handling is the next sector of manufacturing. This chapter deals with the delivery of all the raw materials to fabrications areas, machine loading, the delivery of the fabricated pieces to assembly centers, and the subsequent packaging operations. In John Deere's half-mile-square factory, the computerized system stores over 3,000 parts in each of three high-rise storage areas. The central computer coordinates the production schedule for a tractor under construction in the fabrication and assembly area with the materials-handling system. The computer controls motorized carts that travel around in the storage area, directing them to pull the correct part from the proper high-rise bin and transfer it through the automated plant conveyor system to the proper location for the manufacturing operation. The system is flexible enough to handle virtually any part and can reprogram the materials flow for changing production demands. This system is one of the most complete examples of around a hundred such systems currently in operation in the United States, reducing inventory by two-thirds, thus saving over 10 percent of total production costs. In this sector of automation, the traditional white-collar clerical and scheduling workers are replaced by electronic storing, moving, and controlling systems.

Any time a workpiece and a motion are combined, materials handling is taking place. In fact, it is impossible, in the physical sense, to "handle" anything that is not "material." The definition of robot makes it sound very much like robots are designed with materials handling in mind. The Robot Industries Association defines a robot as "a reprogrammable multifunctional manipulator designed to move material, parts, tools, or specialized devices through variable programmed motions for the performance of a variety of tasks." Notice that the distance of the move is not important. Parts, tools, and specialized devices all fit under the heading of materials, making anything that a robot does materials handling, but a more specific definition of materials handling is required. The common view is that micromoves in a machining operation are not materials handling, nor is the actual spraying of paint, but robots can do such things. In general, materials handling concerns the delivery of items and substances to the scene of the work. For instance, the delivery of the drums of paint to the robot sprayer is materials handling. The distance moved can be large or small, and the material moved can include containers, products, assemblies, parts, tools, fixtures, packing materials, raw materials, finished goods, in-process material, or machines.

A materials-handling task is a task where the robot picks up a piece and puts it down in every cycle. Examples are loading a machine, positioning a part on an assembly, and transferring a part from one place to another. Materials handling means more than just handling materials. More appropriately, materials handling means using the right method to safely move the right amount of the right material in the right orientation and the right condition to the right place at the right time and at the right cost. Robots in materials handling must control virtually all aspects of the transport of the piece. Figure 8.1 shows the standard symbols promoted by the International Standard Organization (ISO) for each of the motions involved in materials handling.

The central issues are the material, the movement, and the method. The type, characteristics, and quantity of material to be moved should be weighed against the source/destination and logistics of the move, the size of one unit of the material, the equipment used to make the move, and the physical restriction on the movement. Like any other engineering project, the design of the materials-handling system should follow the standard six-step design process, as follows:

1. Define the problem
2. Analyze the requirements
3. Generate alternative designs
4. Evaluate and test design alternatives
5. Select the preferred alternative
6. Implement the system

To plan a materials-handling system in support of a manufacturing operation, several more specific rules are useful. The main objective, of course, is to create an environment that produces a high-quality result, but this is an important consideration with all phases. All areas will operate more smoothly if their input materials are of a consistent quality. Plan the flow of materials, equipment, people, and information carefully, and design the layout and the materials-handling system to adjust easily to changes in manufacturing technology, processing sequences, product mix, and production volumes. Keep throughput and timing in mind to reduce the amount of work-in-progress. Control the movement and storage of all materials. If possible, eliminate manual handling within and between workstations. Deliver

Subfunctions of Materials-Handling Motion Functions	
Function Symbol	**Definition**
Rotation	Moving an object from one defined orientation into another defined orientation about an axis passing through a reference point on the object coordinate system. The position of the object reference point remains unaltered.
Swiveling	Moving an object from one predetermined into another predetermined orientation and position through rotation about an axis outside the object.
Translation	Moving an object from one predetermined into another predetermined position by translational motion along a straight line. Orientation of the object remains unaltered.
Orienting	Moving an object from an undefined into a predetermined orientation. The position of the object is not taken into account.
Positioning	Moving an object from an undefined into a predetermined position. The orientation of the object is not taken into account.
Ordering	Moving objects from an undefined into a predetermined orientation and position or direction of motion.
Guiding	Moving objects from one predetermined into another predetermined position along a defined path. Orientation of the objects is defined at each point.
Transferring	Moving objects from one predetermined into another predetermined position along an undefined path. The degree of orientation remains unaltered.
Conveying	Moving objects (or bulk goods) from one random into another random position. The path of motion and orientation of the objects during motion need not necessarily be defined. (See VDI 2411.)

Source: International Standard Organization.

Table 8.1 Functional symbol standard of materials-handling motions

parts and tools to workstations in predetermined quantities. They should be physically positioned, or oriented, to allow automatic tool changing, parts feeding, insertion, and assembly. Use available computing resources to integrate the processing, assembly, inspection, handling, storage, and control of material. Allocate space carefully, utilize overhead space, and consider the effects of machinery on traffic patterns. Figures 8.1 and 8.2 show a materials-handling application.

Elements of the Handling System

All materials-handling systems share the same basic elements: a container, a movement device, an actual trajectory of motion, some type of storage capacity, and a controlling system. In order to design a materials-handling system, all these elements must be taken into consideration. Many products require several containers, motions, and transport devices during the manufacturing process. For instance, an automobile motor may require a special conveyance, but its constituent parts could not be handled in the same way. After it is installed, the motor is handled as part of the car body.

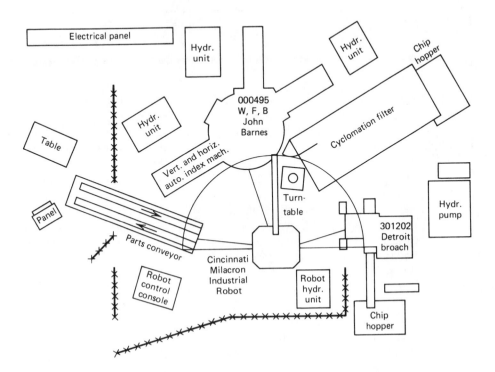

Figure 8.1 Layout diagram for a Cincinnati Milacron T³ 566 robots serving several machines and conveyors

Figure 8.2 Photo of the cell in operation

Materials-Handling Principles

There are 20 principles that form what the College-Industry Council on Materials-Handling Education has endorsed as a complete solution to the design of materials-handling systems.

1. Orientation Principle. Study the problem thoroughly prior to preliminary planning to identify existing methods and problems, physical and economic constraints, and to establish future requirements and goals.

2. Planning Principle. Establish a plan to include basic requirements, desirable options, and the consideration of contingencies for all materials handling and storage activities.

3. Systems Principle. Integrate those handling and storage activities that are economically feasible into a coordinated operating system including receiving, inspection, storage, production, assembly, packaging, warehousing, shipping, and transportation.

4. Unit Load Principle. Handle the product in as large a unit load as practical.

5. Space Utilization Principle. Make effective utilization of all cubic space.

6. Standardization Principle. Standardize handling methods and equipment wherever possible. *(continued)*

7. Ergonomic Principle. Recognize human capabilities and limitations by designing materials-handling equipment and procedures for effective interaction with the people using the system.

8. Energy Principle. Include energy consumption of the materials-handling systems and materials-handling procedures when making comparisons or preparing economic justifications.

9. Ecology Principle. Use materials-handling equipment and procedures that minimize adverse effects on the environment.

10. Mechanization Principle. Mechanize the handling process where feasible to increase efficiency and economy.

11. Flexibility Principle. Use methods and equipment that can perform a variety of tasks under a variety of operation conditions.

12. Simplification Principle. Simplify handling by eliminating, reducing, or combining unnecessary movements and/or equipment.

13. Gravity Principle. Use gravity to move material wherever possible, while respecting limitations concerning safety, product damage, and loss.

14. Safety Principle. Provide safe materials-handling equipment and methods that follow existing safety codes and regulations.

15. Computerization Principle. Consider computerization in materials handling and storage systems for improved material and information control.

16. System Flow Principle. Integrate data flow with the physical material flow in handling and storage.

17. Layout Principle. Prepare an operation sequence and equipment layout for all feasible system solutions, then select the system that best integrates efficiency and effectiveness.

18. Cost Principle. Compare the economic justification of alternative solutions in equipment and methods on the basis of economic effectiveness as measured by expense per unit handled.

19. Maintenance Principle. Prepare a plan for preventive maintenance and scheduled repairs on all materials-handling equipment.

20. Obsolescence Principle. Prepare a long-range and economically sound policy for replacement of obsolete equipment and methods with special consideration to after-tax life-cycle costs.

Containers are usually highly specific in nature. A milk jug differs greatly from a refrigerated truck, although they can both carry milk. A pallet with registration features and slots or holes for piece mounting requires a design stage of its own. Containers should be treated as tools, replaced when defective, and repaired if possible. The specification of containers needs to include a number of parameters, starting with the physical dimensions: the cubic volume, the area footprint on the floor, and the weight empty and loaded. The structure of the container must take the weight and size of the contents into account, of course, but these factors also have an effect on the life span of the container. Some containers can be used only once; they are mounted directly to an important piece during the manufacturing process and destroyed or dismantled afterwards. Other containers require a recycling effort and a means of returning the empty container to the stage in the process where it is loaded.

The protection of the product is a major consideration. If the product and container are to be shipped long distances or stored in a hazardous area, the container must be strong enough to keep the contents from being damaged. The container must also be constructed in a manner that keeps the contents from moving, bouncing against other contents or colliding with the container walls. Materials such as liquids and fluids require a sealed airtight container to prevent contamination of the factory, the environment, and the fluid itself. The modularity of the containers is important if the container is to be loaded into a larger container or hold smaller containers. For this reason, the dimensions, design, and material of the containers should be standardized. Standardization also eases problems when the container is transferred to another department. The storage of empty containers may require that they be collapsible or easily broken down into parts. Containers should be designed to be stacked, and they should have features that enable them to be mounted on conveyors or hung from assembly lines and rails. Tracking and identifying different containers requires some type of labeling. Some types of parts do not require a container; instead, a bracket or frame provides sufficient support, registration, and identification of the part; such parts may be individually moved directly from moving belt, trolley, or carousel conveyors. Containers include boxes, drums, baskets, crates, magazines, pallets, skids, platforms, fixtures, skid boxes, and racks. In order for the automated equipment to recognize a container, something more than a written label is needed. An automatic identification system can be used. These include optical character recognition, hand-held wand bar code scanner, fixed beam bar code scanner, moving beam laser bar code scanner, magnetic code reader, microwave transponder, pattern and shape recognition, and reflective code readers.

Micromoves occur at a workstation, and macromoves are between workstations and between departments. Basically, micromoves are short, and macromoves are long (or longer). Different factors affect each type of move. The factors that affect both micro- and macromoves start with the load size and weight, including that of the container or mounting apparatus. The distance of the move and the frequency of moves per unit of time determine the speed required. The number of loads carried per move and the mix of the load are also important.

Consider the response time of the rest of the system to the delivery of each load, and synchronize it with the cycle time of the move. Design in compensation for the obstacles to the path of the move, and determine whether this move presents an obstacle to other moves. Make the system flexible to account for changes to the requirement for throughput and flow speed. Macromoves are usually accomplished in batches or loads of numerous parts, using trucks, or continuously, using conveyors. Trucks include hand truck, cart, pallet, jack, stacker, pallet truck, platform truck, tractor trailer, mobile crane, automated guided vehicle, counterbalanced lift truck, narrow-aisle straddle truck, sideloader, turret truck, and straddle carrier. Conveyors include chute conveyor, slide rails, flat belt, troughed belt, magnetic belt, roller belt, wheel trolley, slat trolley, chain trolley, bucket trolley, car on track, screw conveyor, vibrating conveyor, vibrating parts feeder, and air film.

Micromoves are considered moves that a human could accomplish with no more than one step, usually handling only one part at a time. Some of the factors that affect micromoves are the reach of the operator, robot, or machine, the limits on access to the workstation, and the interface requirements of the operator with the machine or robot. The actual cycle time at the workstation, the weight, and other characteristics of the part affect micromoves.

Storage of material affects movement in the sense that the storage site is either the beginning of one move or the end of another. Efficient storage can greatly facilitate the movement of work-in-process, tooling, supplies, and fixtures. Planning the storage facility must take the required and available space into account, with special building constraints, such as floor loading and condition, column spacing, and sprinkler and fire codes. The stackability and the stack height of the materials and the load size and weight are important considerations. The need for a central storage facility versus the need for distributed storage can affect the design, as can the need for storage and retrieval of parts in a specific order. Special characteristics, such as flammability and perishability, require special solutions. Automated storage and retrieval systems are graded according to the unit load. The systems include a picking vehicle for large parts or for expansive warehouses. Small parts can be stored on parts carousels or revolving shelving. The configuration of the storage structure for individual parts that do not have containers requires some kind of rack, bin, shelf, or drawer. Some examples are selective one-deep pallet racks, double-deep pallet racks, portable stacking racks, cantilever racks, decked cantilever racks, drive-in racks, drive-through racks, edge-stacked plate racks, pipe racks, pigeonhole racks, case-flow racks, pallet-flow racks, mobile or circulating shelving, mobile racks, and dial index tables.

The control system for the automated handling system can usually be incorporated with existing computers, but the system will be large enough to require additional computer power. Inventory and process control systems, if they are in place, will reduce to some extent the need to enter information again; the same data base can be used in some cases. Try to eliminate redundancies in computer systems by linking the old with the new into one complete system. There are substantial advantages to running an interactive, real-time system over a batch-control system, but the differences are subtle; most available software for these purposes has elements of both. The interactive system is distributed throughout the facility by placing computer consoles at various locations where they can be used by people at the scene of the work, allowing immediate reporting with limited paper-work. The batch system is just that; batches of information are input at one time, gathered, and prepared, usually by dedicated-computer personnel, into data bases that are processed by computer programs to deliver reports. It is easy to see the benefits of the interactive system; it gives continuous tracking of the process rather than the interrupted tracking of batch systems. The importance of this ability varies from factory to factory. The transaction rate, the need for reporting, schedule constraints and priorities, and the required accuracy of the system are the major deciding factors.

The Materials-Handling Robot

By 1983, more than 50 percent of the robots in the U.S. were performing materials-handling tasks. Robots have been applied successfully to loading parts onto pallets and into machines, packing cartons and kits, picking parts from randomly packed bins, loading and sorting parts on conveyors, and storage and retrieval of parts. Experience shows some rules for robots in materials handling.

Minimize the moves if possible, finding the set of the fewest moves that achieves the purpose. Reduce the number of times a piece is picked up and the distances moved. Combine moves and eliminate redundant operations. Once the piece is picked up, its registration should not be lost. Taking the time and spending the money to retain the registered reference makes the system more cost-effective. Where possible, use overhead monorails for materials transfer. Such systems save valuable floor space and provide minimal obstacles to floor traffic. Standardize containers so that they can cross departmental boundaries without requiring reregistration and/or transfer to another container. From a cost viewpoint, moving pieces is an expensive way to store them. Material in transit is the same as in storage, incurring all the same costs plus the cost of the handling machinery.

Robots in materials handling must interface with many other machines, and the synchronization of their respective functions can be done by adjusting the robot and/or the conveyor, or by implementing a new device. Figure 8.3 shows layouts for two such complex materials-handling systems, using several robots and flexible conveyors. For example, in synchronizing a conveyor with a robot, the motion of the robot can be timed to coincide with the delivery of the part, or the conveyor can be fitted with an indexed location that provides a temporary stop for the part. Alternatively, a recirculating or accumulating conveyor can be used, or another robot can be used to pull parts from the conveyor and queue them for the other robot. To retain parts orientation in exchanges between conveyor and robot, a registered pallet or fixtured container can usually be loaded more easily than the individual part, and the same pallet can be used with different fixtures for different parts. Locating guides on the machine tool can greatly reduce the need for precise motions by the robot. A separate device for parts orientation can be used, or the robot can be fitted with vision or sensing apparatus that will enable it to orient the part. The coordination between the robot's execution program and the many diverse parts encountered requires identification of the part. Sensors can be used to indicate the presence of the part at a certain spot. Machine-readable codes on the part or container can be used, or the proper program can be selected, based on information gained through feature recognition by robotic vision or sensing.

Robot Mountings

Stationary pedestals are the most popular type of mounting, because the design and implementation are easy. Robots can also be mounted directly on a machine for local loading tasks. Pedestal-mounted robots are usually placed where they can serve

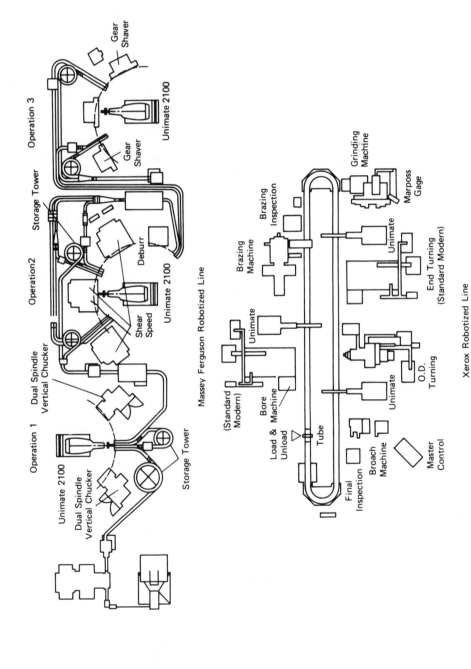

Figure 8.3 Two robotized flexible conveyor lines

two or three machines. In most cases, this is a pedestal mount in a location central to the machines. By 1985, 15 percent of Japanese and 6 percent of U.S. computer numerical control (CNC) machines had some kind of automatic loading systems. Many older machines can be retrofitted with automatic robotic-loading equipment. This type of robot usually picks a part from a registered conveyor or pallet and puts it in the machine tool, taking the finished part and putting it back on the pallet or conveyor. Figure 8.4 shows two machine-mounted robots, and the motions involved in the loading of the machine.

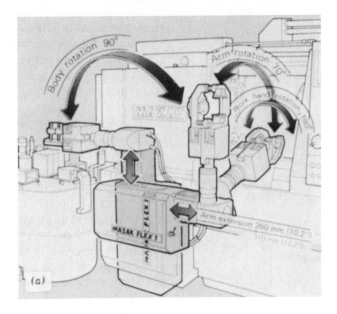

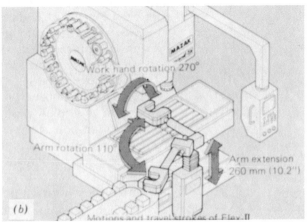

Figure 8.4 Two machine-mounted robots

Robot Mobility

An investment in robotics is expensive and depends on a high utilization factor to provide economic justification. In machining tasks, the robot performs extensive work on the piece. In welding and paint spraying, robots perform extensive work, and the workpieces move past the robot on assembly lines or conveyors. These applications provide a high degree of utilization even when the robot is stationary. A robot that loads a machine tool, however, will be idle much of the time while long machining operations are performed. This idleness is detrimental to cost justification. In order to improve the utilization factor, circular arrangements of machines served by a central robot allow the robot to serve several machines. Problems with access to large machines can eliminate this alternative. Since it is impossible to move the machine tools around, the solution is to move the robot around to increase utilization. The mobility of robots takes three forms, adding one (linear mobility), two (area mobility), or three degrees (space mobility) to the robot's inherent five or six degrees of freedom. Figure 8.5 shows some of the possibilities for installing robots with mobility.

Robots can be mounted on a track to provide linear mobility, enabling the robot to move back and forth on the track to serve several machines in a line. A flat, indexed table can provide motion in two directions, enabling the robot to move back and forth and in and out among machines. This area mobility is also provided by robots on guided carts. Space mobility is provided by a robot mounted from a gantry or on a guided cart with an elevator (much like a forklift). A robot can be mounted almost anywhere to provide mobility. Some tracks are installed on the wall or ceiling to free the floor space. Gantries have been built with two or more robots, and in at least one huge gantry installation, a control cab is actually gantry-mounted, and the operator/observer can move along with the robot. Design of such installations must take into account that as the robot's moves become larger, the cycle time for the routine increases, and the accuracy of each move decreases. The accuracy can be improved by adding registration points on the robot's path that the robot can use to confirm its absolute position. Of course, a stop and a start at the registration point further increases the cycle time. Also, when one robot serves many machines, the queue of work for each machine must be enough to keep the machine busy until the robot returns. This work-in-process inventory can be an important factor in cost justification.

Robots mounted on wire-guided or rail-guided carts can provide virtually unlimited range within a plant. Figure 8.6 shows an automatically guided vehicle with a robot. These carts,

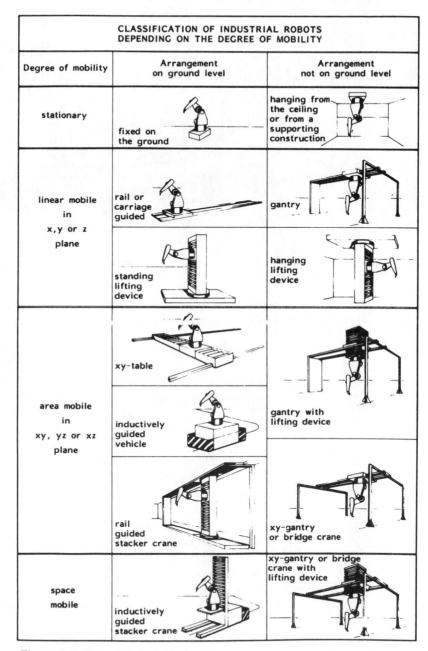

Figure 8.5 Different kinds of robot mobility

many capable of carrying up to 9000 kilograms, do not require a robot to be mounted on them. They carry pallets of parts that can be loaded and unloaded automatically at the workstation, or they can be loaded by local robots. The pallets can double as fixtures

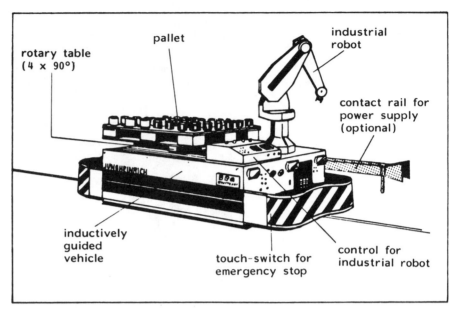

Figure 8.6 Guided vehicle with jointed arm robot

for the parts, as in numerically controlled operations, and they can be loaded and placed in readiness for delivery by the carts. The guiding wires can be hidden beneath the floor, allowing normal walking traffic, and they can be moved easily to adjust to changes in machine layout and cart travel routines. The carts use existing walkways and are fitted with emergency stop sensors in case of contact with an obstacle or a person. Some have sensors that enable them to slow down and stop when they detect a person or obstacle as much as 50 feet ahead. Careful use and effective planning enable such cart systems to reduce warehousing of parts at the machine stations.

Carts can be used to transport a robot to the place of work or to transport raw parts to the place of work and exchange them for finished parts. Mobile robots can be used to handle pieces at the place of work. In all these schemes, mobile systems should be free of the need for manual intervention. They should have their own independent power supply or be able to automatically plug into a power supply at the work place and at various other locations in the plant. They should also have independent controlling apparatus and some means of relaying their position and status to the central control computer. Such mobile systems should be modular in construction to facilitate future changes and to limit costs. The mobility should not interfere with the operation of the process cell, and the exchange of raw for finished pieces should be designed to be automatic and not require that other operations stop. The carts should all be capable of carrying some type of

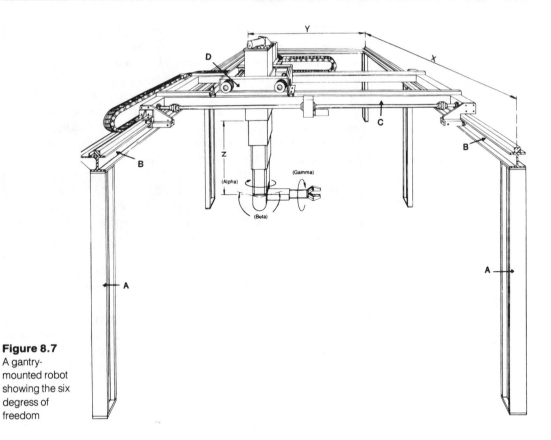

Figure 8.7
A gantry-mounted robot showing the six degress of freedom

universal conveying device or container, such as a pallet. When a robot is mounted on a cart, it is important to make the cart sturdy enough to support all motions and shifts of the center of gravity of the robot as well as all the secondary vibrations and natural frequency effects.

Gantry-mounted robots have the heaviest payloads of all robots and, because of their overhead mounting, can do the work of several stationary robots. The gantry robot takes up a lot of space, with its superstructure, or box frame, roughly defining the work envelope. Figure 8.7 shows a typical gantry robot. The pillars shown as A in the figure define the superstructure by their height and placement. The top side rails of the superstructure are called the runway and usually define the x-axis of the gantry (denoted B in Figure 8.7). The bridge (C) moves along the runway in the x direction. The carriage, denoted as D, moves back and forth along the bridge to define travel in the y direction, and supports the mast or telescoping tubes that travel in the z direction. The robot shown has a total of six degrees of freedom, including the three angular motions shown as alpha, beta, and gamma.

A Gantry robot, in addition to its large work envelope, has the ability to place the end effector in the same spot with different axial coordinates of each joint. These redundant orientations enable the gantry robot to work around obstacles. The gantry can also be lengthened by adding to the superstructure. Huge gantries have been built of up to 170 meters in length (x-axis), 13 meters in width (y-axis), 8 meters in height (z-axis), and lifting up to 15 metric tons. Gantry robots also have some disadvantages. The superstructure is usually larger than the work envelope and the space over the superstructure must be clear to provide room for the z-axis mast. The mast must be able to lift the robot over any machine in the work envelope. The end effector, if required to lift heavy loads, must be very strong, incurring a weight penalty that affects some velocity-accuracy trade-offs. Also, the torque-weight considerations for the angular motion of the robot arm are the same as when the robot is on the floor; the gantry achieves the higher payload in the x-, y-, and z-axes, not in the alpha, beta, and gamma rotations. The program must take this into account, making heavy-payload moves in straight lines, not arcs. Safety in the work envelope is complicated by the fact that the overhead installation of the robot takes it out of the normal view of people. Separate areas inside the work envelope must be locked out from the robot to allow service on one machine while the rest of the machines are working.

Parts Preparation for Robot Handling

Jigs and fixtures are used to enable the robot to pick up parts in the same attitude and position and are a cheap way to give the robot a registration every time it picks a part up. Robotic vision systems enable the robot to pick up parts that are not located precisely. In handling arbitrary shapes, such as rough-hewn billets, the gripper and fixture should be of a configuration that is immune to, or automatically adjustable for, the arbitrary features. Robots are often required to pick up a part after partial machining, put it down, and then pick it up in a different orientation for completion of the machining operation. Grippers and fixtures in these applications must be able to handle the piece in its partially completed form and maintain the registration. Reusable and programmable fixtures can be designed to handle complete parts families. In applications where there is a mix of parts, some type of identification of the different parts is needed. Laser coding systems can be read by the robot but require that the pieces be labeled. This can be impossible if a part is to be machined in rotation. Vision systems and sensory apparatus can be used to recognize significant features of a part, such as diameter or shape, in order to use the correct machining program.

Different weights of parts should be taken into account in the program, so that the accuracy of the move is optimized. The different parts may require adjustable grippers or the ability of the robot to exchange grippers.

Integrated Systems

Over the years, manufacturing has adopted many elements of automation technology. Numerically controlled lathes, mills, and routers are common today. When the first numerical control (NC) lathe was installed, it brought with it two boundaries between the automated and the nonautomated: the input boundary and the output boundary. Something nonautomated brought workpieces to the lathe and something nonautomated took it away. This lathe became an island of automation in a nonautomated environment. Most of the problems with automated systems occur at the boundaries that they have with nonautomated systems. The machines in the system have been designed and tested to work together, to generate only a certain number of internal situations, and to be able to handle each one. Such a system depends greatly on external factors, of course, to be applied to a task. Planning the external factors is a concern of implementation and installation. The lathe is prepared to do a job, but it requires that the workpiece be installed correctly and that it be of a consistent quality and shape. When the system is installed, it must be set, or programmed, to accept goods of a certain type and in a certain physical orientation. Once the workpiece meets these conditions and the lathe completes its operation, loss of the orientation causes expensive reorientation when the piece moves to the next phase in the manufacturing process.

From a systems viewpoint, each island that requires reorientation because an upstream island destroyed that orientation injects an additional cost factor. Each island may be saving enough money to make a net gain, but some of the savings are lost. The selective application of automation is a necessary step, but it should be considered a temporary aspect of a phased plan to integrate all the islands of automation into an integrated factory system. The islands of automation must be tied together and synchronized. The materials-handling system works with the central control system to build the physical bridges between the islands and get the information required to coordinate activities.

Selection of the specific islands to be linked must address the cost, quality, and cycle-time benefits; in other words, the same criteria that were used to decide on the implementation of the original automated system. The islands can be as small as a single workstation or as large as a whole department. A good place to start is the central computer, after the computer is installed, and all the numerical control machine tools can be scheduled centrally, and all the machine centers and parts-loading robots can be tied to visual inspection systems. Tie together all the robots for welding, assembly, painting, and inspection, all the lasers for cutting and finishing, the automated storage/retrieval systems for storing work-in-process, tooling and supplies, smart carts, monorails, and the conveyors for moving material from station to station.

CHAPTER
9

Flexible Manufacturing Systems

This chapter discusses the uses of robots in a flexible manufacturing system. Because robots are ideally suited for applications in manufacturing environments, their role in the successful implementation of a flexible system is very important. This chapter describes what a flexible system is, uses for flexible systems, planning for implementing a flexible system, economic issues of a system, and other topics related to robotics being used in a flexible manufacturing system.

Background

Manufacturing experts generally agree that if the United States' manufacturing industry is to have a prosperous future, especially when facing increased manufacturing competition from abroad, automation and flexible automation is not just a necessity, it is a priority. To regain our status in the world industrial community, the U.S. must excel in the manufacture of durable goods. We must evaluate our current manufacturing methods and, following the lead of our overseas counterparts, begin to automate our predominately outdated manufacturing methods.

Our manufacturing methods have evolved over time from the predominantly manual methods of yesterday to the automated systems that are being developed and implemented today. Industry has sought to automate for obvious reasons: increased production capacity, improved quality of goods produced, and safety. As a result, machines have been developed that can cut, drill, forge, and so on. Predictably, these machines are vast improvements over the manual methods of production, yet there are still many instances where automation has proved to be inefficient. The human factor still dominates. The most powerful and efficient lathe is useless without a competent human operator, optimum scheduling, and proper planning.

Studies have shown that production equipment in U.S. industry is vastly underutilized. This is directly attributed to the amount of time production equipment sits idle, without being scheduled for work. Even when work is scheduled, there are serious problems, because value-added time is seldom achieved.

Value-added time refers to production time that can be attained through efficient equipment scheduling. It usually involves second and even third operating shifts and other methods that reduce equipment downtime and promote equipment utilization. Inefficient capital equipment utilization invariably leads to another problem, because inventories of materials and supplies must be stored while waiting to be processed. In many manufacturing endeavors, it is not uncommon for millions of dollars of inventory to be stored, sometimes for months. This ties up the company's capital (as well as warehouse space) and produces no return on investment during the storage period. It is also not unusual for the value of the materials inventories to vastly exceed the actual cost of the company's capital equipment. Unscheduled equipment time and inventory storage are capital drains on any company, but it is especially damaging to companies that do not have large financial reserves. The possible capital gains that can be achieved from proper equipment scheduling and use exceed any method of reducing direct labor costs.

It is easy to see that proper equipment scheduling leads to frequent turnover in materials inventories, and not coincidentally, reduced lead time for tooling up new applications and improved response to consumer demand. These are important factors that contribute to overall manufacturing financial success and viability. Robotics can play a very important long-term role in improving manufacturing equipment utilization. But robotics is only a part of the solution. A complete system for automating a manufacturing plant is what is needed. Here, too, robots play an important part in developing an efficient, productive factory of the future. The increased momentum to implement robots and other programmable automation devices is an important factor in increasing the manufacturing output of the U.S. The overall role that robots will play in automation can be ascertained by examining the concept of using robots in a flexible manufacturing system.

Flexible Manufacturing Systems: An Overview

A flexible manufacturing system (FMS) is an arrangement of machine tools that is capable of performing industrial manufacturing work—grinding, drilling, welding, cutting, painting, etc.—in a stand-alone environment while being tended by one or more industrial robots. Stand alone refers to the grouping of the machining tools in various workcells to produce a complete manufacturing arrangement. The machining tools are interconnected by a workpiece transport system. Also included is a secondary transport system incorporating one or more robots. Each robot is assigned the task of moving workpieces to the machines from pallets or other delivery apparatus. The robots are also responsible for moving workpieces between machining stations and out of the system upon completion of machining.

The entire system is controlled by computer programs designed specifically for robotic applications. Usually, a central computer processes all instructions to the system and continually monitors the system during operations. For larger, more complex systems, two central processors may be necessary. Because of the versatility of computer automation, a flexible manufacturing system can have a variety of parts being processed at one time. The computer controls the systems robot, transport mechanisms, and the individual tooling or processing machines.

The physical make-up of an FMS usually includes one or more machines that process a selected variety of workpieces without requiring human operator attention for parts changes or the alteration of machine cycles to accommodate different machining applications. To perform this task, a flexible manufacturing system can be comprised of one or several workcells. A cell is a manufacturing unit that is usually made up of two or more workstations or machines and the material transport mechanisms and storage buffers that interconnect them. A typical workcell is shown in Figure 9.1. Storage buffers are areas in the system where raw materials are stored before processing, and where workpieces are unloaded and stored between machining operations.

Figure 9.1 A typical layout of a single-machine system

Simplifying the entire process, a flexible manufacturing system operates under computer control. In a typical FMS configuration comprised of two or more workstations or cells, different workpieces are placed on pallets or some other materials-storage medium and situated in front of, or within short proximity to, the processing machine. Each workpiece is indexed for the machining operation it will undergo. Indexing is a presorting process whereby the materials on a pallet are arranged in such a manner that they can be retrieved by robot arms fixed with grippers in the order necessary for machining. Indexing also ensures that the materials to undergo machining are properly registered. Workpieces are arranged in a predefined location, established by the location of geometric points on an x-y plane. Through control of the software, the robot arm moves to the exact location of the needed indexed workpiece, grasps it, and moves it to the next predefined location—the actual processing machine or tool. Again, the registration point is predefined. The geometric coordinates where the piece is to be loaded are transmitted from the computer to the robot. Automatically, the arm places the piece in the proper position on the tool for processing, and the machine processes the piece based upon the instructions provided by the software. When the first tooling is finished, the arm grasps the piece and moves it to another processing cell. When machining is complete, the workpiece is is unloaded and, in some applications, gauged for precision. Ultimately, it is placed on an exit conveyance for transport to a storage or packaging area. A typical assembly cell containing four robots is illustrated in Figure 9.2. The actual work process, including the flow of work through the system, is shown in Figure 9.3.

The Role of Robotics in an FMS

The use of industrial robots as the center of a flexible manufacturing system entails far more than using robots for simple materials-handling and pick-and-place operations. Combining the programmable task-mastering capabilities of the robot with automatic controls that are provided by numerical control (NC), computer numerical control (CNC), and direct numerical control (DNC) machine tools, a manufacturing system can be developed that provides industry a high degree of true flexibility and a greater potential for productivity. Robots play an important part in the successful flexible manufacturing system. In fact, flexible systems would probably not exist without robots. Their capacity for efficiently performing machine loading/unloading tasks make them ideal for automated manufacturing methods. Aside from their capability to perform automated loading/unloading tasks, robots increase product quality and manufacturing control.

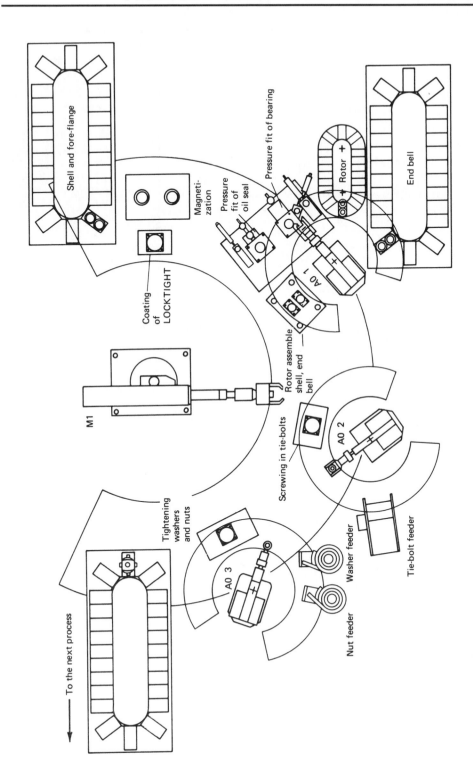

Figure 9.2 A flexible system for assembly

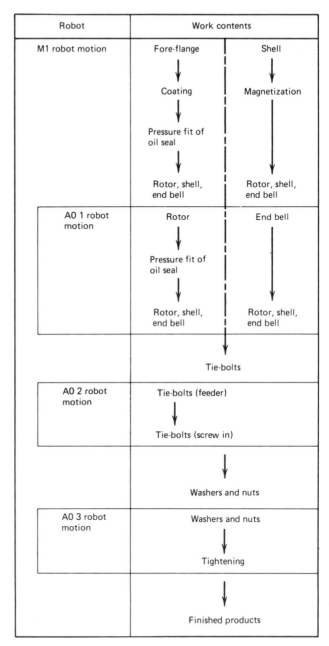

Figure 9.3 Work flow through a flexible system

The use of the robot's manipulative and transport capabilities for loading/unloading workpieces is generally more sophisticated than for simple materials-handling functions. This is due to the evolutionary development of robots themselves.

Early robots were developed for simple pick-and-place applications. The technology was not available for more sophisticated functions. Today's robots can grasp a workpiece from a conveyor belt, lift it to a machine, orient it correctly, and insert or place it on the machine for grinding, drilling, cutting, etc. After processing, the robot unloads the workpiece and transfers it to another machine or conveyor. Even while machining takes place, the robot can perform another task until the computer signals it to unload the workpiece.

Current FMSs that incorporate robotic applications include the loading and unloading of hot billets into forging presses, machining applications using lathes, welding, stamping presses, and plastic injection molding. The automobile industry has developed many systems featuring robots for parts assembly, parts transfer, painting, sealing, deburring, palletizing/-depalletizing, and applying adhesives.

Few flexible manufacturing systems today are truly flexible. Most systems have limited functionality and limited part-number mix (often as low as two or three). Many systems produce one type of parts or are very limited in diverse part-machining applications. The primary reason for this is the fact that true flexibility requires a large number of tools for machining a variety of workpieces and central computers that are capable of instantly processing sophisticated FMS software. Most flexible systems have limited capabilities as far as their capability to automatically handle a variety of tools and workpieces. Applications are growing, but slowly and with much up-front expense. Exclusive use of robots in FMSs will undoubtedly require that the robots provide for the rapid interchanging of tools, gripping devices, and parts with minimum disruption of production.

Why Robots?

The primary motivation for using robots in a flexible manufacturing system is to reduce direct labor costs and increase productivity. Manufacturing efficiency can be achieved even when only a single robot is used for small-scale manufacturing tasks, because a robot can do more than perform a single task. Simply by reprogramming, a single robot can be taught to perform many different functions, each of them efficiently. For example, a single robot may be used primarily to service several different machines comprising a flexible system. Figure 9.4 shows a single robot servicing several lathes. Rather than letting the robot sit idle while the machines drill, cut, deburr, or whatever, it performs other operations, such as retrieving the next needed workpiece and placing it in a staging area, while waiting for the machining tasks to be completed. This type of efficiency naturally leads to more productivity.

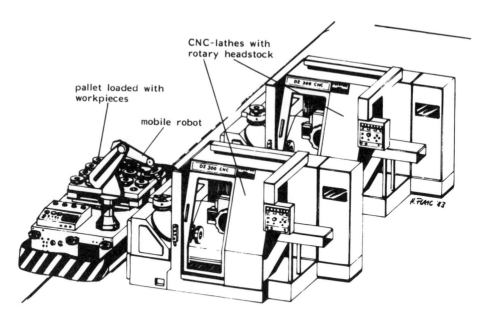

Figure 9.4 A robot servicing several lathes

True efficiency results when the components of an FMS are utilized to their utmost. In other words, rather than have a single robot that loads and unloads parts into a machine designed for drilling, it is much more advantageous to implement a system where multiple operations occur with limited production interruption. In applications where one machine is capable of several different tasks (drilling, cutting, deburring, etc.), production improvement is as easy as performing a tool change. Robot arms are capable of performing tool changes; the computer tells the robot when and what tool to change. A system of this type eliminates the need to have a workpiece transported throughout a factory so different machining tasks can occur. It is the transport and waiting that eats up efficiency. An arrangement of this type keeps the machinery busy and operating continuously.

Industrial applications using robotics in flexible systems show significant savings in direct labor rates. A cost savings of 50 percent to 75 percent is not unusual in a well-designed and utilized FMS. Not only are robots less expensive to operate and maintain than their human counterparts, they are more efficient and productive. Robots don't need vacations, breaks, pay raises, benefits, etc., and they have the capacity to work 24 hours a day, 365 days a year. Flexible systems featuring robots combined with the maximum equipment utilization are profit builders, not profit eaters.

What is Flexible?

In the sense of robotics, flexibility refers to the fact that the robot is adaptable and capable of being redirected, trained, or used for new purposes. In an FMS, flexibility refers to the reprogrammability or multitask capability of a robot or machining tool. An industrial robot helps make a manufacturing system flexible. A robot obtains its flexibility through computer programming and its capability for performing different tasks. The development of robots has coincided with the development of CNC machines. The ability to reprogram these machines makes them capable of performing different functions together. Both are controlled by the computer software, and the software instructs them how to perform tasks together.

Consider a typical CNC machine. The computer program instructs the machine how deep to drill, how much to cut, and where to weld. Everything that the machine does is controlled by the computer. A computer program that instructs an automated press to drill a workpiece to 3/8 of an inch can be modified quickly to drill to 1/4 of an inch instead. A change of this type might take a good CNC programmer five to ten minutes to accomplish. A robot arm is just as easy to reprogram. This is why a flexible system is flexible. FMS computer programs can be modified quickly to allow for different sizes of workpieces, different machining specifications, different tools, adjustments to transport mechanisms, and a variety of other industrial applications. The computer programs that control the operation of an FMS are written with production efficiency in mind. The system programmers work with manufacturing engineers to design optimum programs that promote the maximum utilization of equipment.

If a machine tool is capable of performing both cutting and drilling tasks, the software is written so that the machine can be used to perform both tasks. This same effort applies to the programming of the robot that tends the machine. If a robot is capable of multitasking, the program controlling the robot should reflect this fact. This ensures maximum utilization of the robot's capabilities. Programming of flexible systems increases machine utilization, which in turn increases productivity. Idle time is eliminated. However, the real value of flexibility is seen in the retooling of machines. Whenever a machining process must be retooled, for whatever reason, machinery often sits idle. Time is wasted waiting for the new tooling specifications to be designed and tested, so that the new procedures can be taught to the machinist. Reprogramming eliminates this costly process. If a manufacturer must change a part's design to meet safety criteria, as soon as the design change is complete, the computer program is modified. Once the program is modified, the new part is produced, meeting the new specification. Almost overnight a retooling effort can be accomplished. New products are a bit more complicated, but the effort is essentially the same.

The Need for Automation Flexibility

Long before robots there was manufacturing. However, while robots have improved, for the most part, manufacturing methods have not. Robotic technology has grown at a faster rate than technological improvement in industry. Current production methods in the U.S. use little of the automation that is available. From a pure business standpoint, if American industry were more conscious of the economic impact of technology, it might be more inclined to improve antiquated manufacturing methods and perceptions. American industry needs automation. Flexible systems are a part of the automation process. Flexible systems can produce more durable goods of higher quality than those produced from other methods. Experts generally agree that the implementation of an automated flexible system for manufacturing could result in a significant reduction in the real cost for capital goods. This could produce drastic changes in the prices of manufactured goods throughout the economy. These cost changes would in turn have a dramatic impact on the rate of price inflation. Supply could truly serve demand.

Flexible systems provide the potential to produce a range of different products in large numbers. Conventional systems produce a wide range of customized products with low levels of output for each product. For centuries, this type of job-shop processing has produced many products. Such manufacturing is essentially batch processing. Manufacturing by batch processing covers the range between the extremes of one-of-a-kind products to the mass production of standardized products. The same part can be reproduced in the millions, but the cost-to-product ratio remains high. Unit costs in these shops have always been significantly higher than the unit costs of products that are mass-produced in flexible arrangements combining specialized production equipment, such as that employed in the automobile industry.

The biggest obstacle in batch processing is lack of flexibility. In batch manufacturing, the manufacturing equipment is not readily adaptable so that a new part (of the same type) can be produced. The effort to make a changeover to produce the new part directly governs the cost and production of the part. If it costs too much to change, it is not economical. If it takes too long to change, it is not economical. Presently, it is very expensive for a manufacturer to make a change from from one batch process to another, no matter how similiar. A flexible system does not have this problem.

Flexible systems also have a direct impact on the reduction in labor costs. Existing and future flexible systems are likely to replace significant numbers of machine operators and other unskilled laborers. The replacement of these workers by automated flexible systems will decrease direct labor input. Studies by Carnegie-Mellon University estimate that in the factory of the future about 75 percent of the unskilled laborers could be replaced by robots in flexible systems. The cost savings to industry could be very dramatic. Today, flexible systems have managed to reduce some labor costs by as much as 20 percent, depending upon the industry and the application.

Perhaps more significant than impact on direct labor costs is the impact that flexible systems have upon manufacturing capacity. True flexible systems can work around the clock. Naturally, production capacity increases with maximum machine utilization. A well-designed FMS is an efficient work unit, regardless of the application. Flexible workstations are numerically controlled, leading to a high degree of production predictability. This results in a short lead time, lower in-process inventory, orderly scheduling, and more accurate predictions regarding the level of product quality. One area that flexible systems will significantly impact with increased cost is in materials inventories. Flexible systems usually do not reduce levels of materials inventories. In fact, inventories often need to be increased, because of the higher level of machine utilization (accompanied by higher output rates). There must be adequate reserves of materials, so the machines can be kept busy.

FMS Planning

Regardless of the FMS type, complexity, or application, all flexible systems have similar requirements that must be met before successful implementation is possible. These requirements relate to justification for detailed advanced planning and coordination. Even with a great degree of flexibility, robot installations must be designed around the application, using a complete plan that encompasses the total production strategy. Planning is the most important phase of any project, and robots are no exception. A robot installation can be extensive and expensive, requiring an effort equivalent to a massive retooling. The entire factory must reflect a strategy of accuracy and efficiency to take full advantage of the robots.

Planning for a robot installation begins with identifying the tasks to be done. Tasks that can be easily isolated from the rest of the process are the leading candidates for robotizing. The biggest problems encountered in robotizing occur at the boundaries of the robot's task. Pick a task in the process where the workpieces are already organized or registered. Carefully analyze the potential robot's requirements for input and the condition required of the output. Complete robot workcells have been designed to take a workpiece from one stage to the next, but problems can occur getting the workpiece to the workcell and then getting the workpiece out when the task completed.

Some of the problems associated with robot systems can be traced to things outside the robot's control. Incoming workpieces must be organized on a pallet or conveyor before the robot can be held responsible. The entire production process must be geared toward setting things up for the robots and accepting output in a similar format. This is why the whole process must be examined. A robot workcell that welds car body panels onto chassis will fail if the chassis is upside down or even one inch from the expected location. A human can see this and compensate, but that ability in a robot is very expensive. The alternative is to mount the chassis on a fixed platform in a known location. The registered location is where the robot expects to find the workpiece. If the robot knows the chassis is upside down, it can weld successfully. A successful robot can identify weaknesses in the basic production strategy. If the robot completes work before the downstream resources are available, the result is a bottleneck or a storage problem. If the robot is ready for work before the upstream processes can deliver, the full benefit of the robot is not being realized.

Determining the extent of the robot system is crucial. It is easy to go overboard, because the abilities of modern robots are so advanced. The common wisdom is "all or small," meaning that a successful robot installation is either so comprehensive that interface between robot and human is limited, or so small that the interface is simple. At one extreme is a system that accepts iron ore and delivers sheet steel. At the other end is a robot that takes sheet steel from one pallet and places formed metal on a stack. One requires a whole system built around robots, the other requires only a loaded pallet. Unwillingness to completely restructure the entire factory has placed most manufacturing concerns in the position of being able to implement only limited robotics projects. The most routine tasks are naturally the first to be robotized. Automobile assembly lines provide a natural register for the workpiece auto chassis, and robots have shown perhaps their best success in spray painting and welding car bodies. Routine tasks are easy to precisely define but are the most difficult tasks for humans to accurately repeat.

An important consideration when planning a flexible system is the scheduling of workpieces through the system. The aim is to minimize any disruptions caused by tool changes that require manual intervention and the maximization of machine/robot utilization. The successful implementation of robots in a flexible system requires situations that incorporate a large volume processing and a high degree of uniformity. Robots are designed for such tasks. They bring to this environment supreme versatility due to their design for performing multiple-repetitive tasks at fixed expense and their inherent degree of task uniformity. Low-volume assignments are not best suited for robots although robots can perform low-volume tasks with the same degree of uniformity as for high-volume applications.

When planning a flexible manufacturing system, a design engineering team must consider a number of important functions and capabilities. The system must have priority machine selection capabilities. This means that all robots to be used in the system should be programmed to accept signals only from those machine tools that are not already busy with a task. In applications where several machines send signals to a single robot, a priority decision-making routine must be installed, or chaos results. The robot has no way of knowing which machine to service and in what order. If any machines in a flexible cell are not operating, the robot should be programmed to recognize and serve the machines that are operating. When a down machine is placed back in service, the robot can recognize this and serve the machine as required. With an arrangement of this type, no human operator intervention is required to keep the system producing parts.

The selection of workpieces for an FMS should target the most efficient processes for task performance with regard to the size, tooling, range, and weight constraints of the entire system. It would be futile to try to incorporate a robot into an FMS to handle parts weighing several hundred pounds when the robot can only handle a part weighing less than 100 pounds. One of the most important considerations in an FMS is the inclusion of an on-line inspection operation. This allows parts inspection to be kept in close proximity to the system or cell. Naturally, this saves time and promotes better control of the manufacturing process. To illustrate this, a robot can be programmed to off-load parts onto a gauge station where checks are performed to check parts dimensioning. If any part is out of tolerance, this information is transmitted to the central computer so adjustments to the tooling machine can be made via on-line programming capabilities. If parts are far out of tolerance, robots can be programmed to reject the part and transfer it to the appropriate receptacle. This high degree of flexibility through programming ensures the continued production of quality parts throughout the system without slowing down production.

Another consideration is how parts and workpieces will be brought in and taken out of the FMS. It is important that the software specify the exact location of workpieces or materials so that the robot(s) know where to go to retrieve them. Coincidentally, it is also important that the workpieces or materials to be machined be placed in the correct position specified by the software. The software should also include the location specifications where finished parts are to be placed for removal from the FMS. Buffer storage must be included in the system. Buffer storage ensures that a system does not come to a complete halt if one of the machine tools goes out of service. For example, should a cell become inoperative because of lack of parts, machine tool failure, or other problems, the storage buffer provides that the entire system does not need to shut-down, as parts can be stored until the machine is brought on-line or retooled.

An FMS should include the capability to palletize finished parts after inspection and completion. In this case, the robot should be programmed so that the finished parts can be automatically stacked to the desired level and any packing materials added. Programming the robot for this task keeps the entire line operating without having completed parts stacking up waiting to be manually removed.

For maximum utilization and productivity, robots should be capable of handling more than one part at a time. This is particularly important in machine applications involving machine-loading functions. Robots that are capable of picking up two or more parts at one time and loading them into chucks or collets perform more efficiently and productively than robots that handle a single part at a time. Programming flexibility allows for multiple parts-handling requirements. FMS software also needs to address any factors where changeover of tooling or operation modifications occur. Should a part size or configuration be changed, or if machining tasks need to be added or deleted to machine a particular piece, the software should be readily updatable so that the robots can be reprogrammed quickly to accomodate the changes. It is also possible to use the robots themselves to accomplish some tool changing tasks via the software.

Overall, for the first-time user, the most likely scenario is not so grandiose. Make the initial installation easy to engineer and install, with a moderate payback. Steer clear of an initial installation in a crucial area. Failure in a high-visibility operation can put a damper on the whole program. Rank the potential applications according to complexity of installation, possible savings, impact on labor, capacity gain, and quality improvements. Analyze the potential match-ups of applications with the most feasible robot system. The most extravagant robots may be overkill for your application. Segment the task to see if any parts can be completed with simple robots. Old-fashioned tool engineering can be an effective complement to the more advanced robots, allowing some tasks to be done with more available standard machinery.

Training is a must for the people who will be interacting with the robot. It must cover aspects of maintenance as well as start-up procedures and operation. The safety of these persons is a function of their knowledge of the system. Dangerous areas should be clearly marked and preferably barricaded to prevent even accidental entry.

Other criteria to be included in the design and implementation of an FMS is:

1. The number of different machining processes necessary to achieve maximum utilization
2. The number of manufacturing cells necessary
3. The robot's rate of work
4. The number of parts needing machining
5. The tolerance of parts needing machining
6. The speed of machine processing
7. The conveyance methods for materials transport
8. The number of shifts needed to produce the required number of parts
9. A consideration of the raw materials to be processed (dimensions and weight of each piece)
10. Planning for proper materials flow
11. Planning for each manufacturing cell
12. Integrating machine tools, robots, workpiece carriers, and overall system control
13. The organization of the entire system for efficiency and cost-effectiveness
14. Scheduling of workpieces through the system
15. Programming processes designed for maximum flexibility
16. The amount of human supervision required

Humans and the FMS

When robots are used in a flexible system, the entire system is usually jointly supervised by a computer and humans. Human controllers sit at a computer terminal and oversee the automated processing functions. Though designed to be minimal, the human factor is nonetheless important for the overall efficient

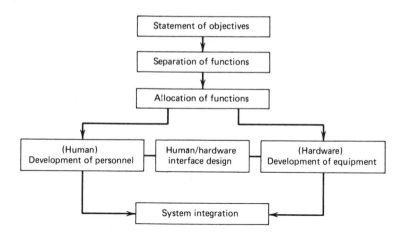

Figure 9.5 Allocation of functions—human and machine—in an FMS

operation of the FMS (Figure 9.5). The human component is an integral part of the entire automated process, and the human operator should be in a position of control. Operators should have an understanding of the software controlling the system as well as the operation of the controlling computer. Operators possessing competency in numerical control programming are particularly desirable. The ability to override computer control is also important, because a human operator is much more flexible and adaptable to any unique situations that may arise. The human role is coherent with that of the machines and must be planned during the design stage. The main reason for the coherency is the psychological needs of humans. If an FMS operator is given too little to do, boredom results. If the operator is required to perform a myriad of supervisory tasks, mental overload may arise. Both factors lead to poor performance by the human and less productivity and efficiency. Strive for the level between the two extremes. Enriching job experiences often provide less fatigue and psychological stress than simplified, mindless routines, because simplified tasks have fewer decision-making opportunities.

When allocating system responsibilities between human and machine in an FMS, be cognizant of levels of interest for the human and the degree of decision-making responsibilities between human and machine. The tasks that the human performs and the tasks that the computer performs are separated. There can be some overlap in assigning these functions, as this promotes reliability and efficiency in the system. An operator in an FMS may be required to shift attention among a variety of different machines, making sure that each operates properly and is not in need of service or maintenance. Humans tend to be more responsible for this type of activity featuring multiple, diverse tasks.

FMS Applications

There are many applications for using robots in a flexible manufacturing system. While it would be impossible to list them all, a selection is given here. Each application is outlined according to the manufacturing operation and its associated robotic applications.

Materials Handling:
Moving parts from a warehouse to machines
Depalletizing
Transporting explosives
Stacking parts
Loading from roller conveyors to monorail
Core handling

Machine Loading/Unloading:
Loading parts for machining
Loading parts onto CNC lathes
Loading hot-form presses
Loading a punch press
Loading die-cast machines
Loading plastic injection-molding machines
Loading an electronic beam welder

Spray Painting:
Painting aircraft
Painting automobiles
Applying undercoatings
Applying thermal material to rockets
Painting appliances

Welding:
Spot welding auto bodies
Braze alloying of aircraft seams
Arc welding of auto axels

Machining:
Drilling panels on aircraft
Deburring auto parts
Sanding furniture

Assembly:
  Riveting small assemblies
  Drilling and fastening metal panels
  Inserting and fastening screws

Computer-Integrated Manufacturing (CIM)

Computer automation is being brought into every sector of manufacturing, but the paper shuffling needed to run the entire manufacturing process is still the most significant barrier to a revolutionary increase in productivity. This barrier could, in theory, be overcome with the implementation of full-scale computer-integrated manufacturing, in which a central computer replaces people and paper. However, a central computer that could perform such a tasks would also have to embody the human understanding of how a factory system actually works, being especially cognizant of the most efficient ways to run the factory, and it would have to be programmed to make decisions about what to do in any situation. Because running a factory can at times be a shoot-from-the-hip activity, nobody really knows logically how to coordinate all of the machines, materials, and people, and their disruptions, breakdowns, and bottlenecks.

If the understanding of how a factory works could be put into a central computer, it could respond to and deal with breakdowns and disruptions more quickly than any person. Manufacturing efficiency could then take its great leap forward to true computer-integrated manufacturing. CIM is a new technology that involves industrial and cultural changes in structure. There is a great deal of development work that must be done before CIM becomes a functional reality. Developing CIM depends upon getting a better understanding of how a factory could optimally work and how that ideal could be translated through software into a central computer-controlled system. CIM, if it is to work, might be the technological edge that is needed to boost productivity and make the United States industry competitive. Today there are only a few examples of CIM, representing little more than 5 or 10 percent of what is possible. While there appears to be a trend toward computer automation and robotic installations, American industry has barely begun to invest in the basic technology that is needed.

10

Economics

The objective of this chapter is to discuss the economic issues involved in automating our factories. It would be presumptuous to assume that all of the economic reasons to automate could be included in a single chapter, but the principles that are discussed provide a foundation for any manager or finance officer who may be considering using automation.

Why Automate?

The reason to automate is purely economic. It just makes good business sense to do so. Automation, in spite of the negative press about worker displacement, is an economic necessity today. Automating industry means that goods can be produced faster, cheaper, and without loss of quality. These three reasons translate into profits. The manufacturing functions of a company are either a competitive edge or a corporate liability. There is a distinct connection between manufacturing success and corporate success. Corporate success is a matter of economics, and economic success translates into profits. Computer automation helps achieve economic success. For all of the arguments that are made in favor of automating the factory or against automating the factory, the only rhetoric that financial decision makers should concern themselves with is the economic issue of automation. How much is it going to cost and what benefits can be gained from the investment? This chapter will try to provide some additional insight.

Background

There is a wide diffusion of robots in U.S. manufacturing plants. U.S. companies who are using these automated machines include large, well-known manufacturers and small, lesser-known companies. Surprisingly, more small companies are now investing in the technology, in spite of the fact that it was the larger corporations that first implemented robots for manufacturing applications. Just as suprising is the fact that many companies owning robots are using them for batch and custom production, not just for mass-production purposes.

Prior to 1976, the primary purchasers of robots were large manufacturing companies that employed over 1,000 workers and specialized in mass-production operations. Chrysler, Ford, Gen-

189

eral Motors, and American Motors are included in this group. Small companies, employing under 500 workers, accounted for about 10 percent of the robots purchased during this time. After 1976, 30 percent fewer large companies were making investments in robot applications, while more small companies were purchasing the new machines. Companies engaged in mass production were still buying, but the number of buyers in batch and custom production was on the increase. Those making custom products included a significant number of aerospace companies. By 1981, the number of companies buying robots was just about 50-50, small and large. Today, smaller companies and firms involved in producing custom and small batch goods are making the majority of the purchases. This is due to the reduction in costs and the wider applicability of the newer robots. However, in 1981, 30 percent of all robots were owned by just six companies. The number of robots being purchased continued to increase, but those making the purchases were often previous users buying more equipment.

This is starting to change. Robots are not the novelty they were 10 years ago. More companies are buying robots to provide them with the opportunity to test actual capabilities. These companies often wind up investing in additional robots once they see actual on-the-job performance. A company purchasing a robot for evaluation purposes might expect to spend $50,000 to $100,000, approximately the cost of a moderately sophisticated computer-aided design system. Undoubtedly the user base will expand in the coming years, as will applications for robots. New technologies will make robots less expensive and more versatile, thus creating more markets and a larger group of users.

Currently, the users putting the most robots to work are in the metalworking sector of manufacturing. These companies account for nearly 90 percent of all robot users in America. Metalworking involves the design, fabrication, and assembly of products or components made from raw materials made of metal. Manufacturers in this group include companies that produce fabricated metal products, machinery (electrical and non-electrical), and transportation equipment.

This sector of manufacturing has special significance, because it is responsible for the production of all of the tools and equipment that is purchased by other manufacturers within the sector. Its workers and craftspeople are probably the most experienced and knowledgeable people in manufacturing. The process is refined and efficient. This sector has always been one to make the most of new technology and has never been afraid to experiment with new ideas or techniques. Its knowledge base and experience is widely used in many other areas of U.S. industry. It is also the sector employing the most robots where significant cost savings have been documented.

Flexibility Versus Efficiency

Manufacturing companies are usually organized according to the goods being produced and the size and length of their production runs. In general, the wider the variety of products or parts made, the smaller the production batch. A manufacturer's ability to design, tool up, and produce goods in a minimum amount of time is directly influenced by the number of different products that must be produced overall. Logically, it is more efficient and economical to produce as many goods as possible once a production configuration is established; however, in reality, this is not always the case. It would not do a manufacturer a lot of good to make 10 million brake drums when only 2 million could be sold. It is more sensible to produce the quantity marketable and spend other time manufacturing other products that have market potential.

In traditional manufacturing, the changeover that is required when switching from one product to another creates inefficiencies. Manufacturers often cut batch sizes, so they have ample time to reconfigure equipment and processes in order to make the changeover to another type of product. The time spent in reconfiguration is lost time, since products are not being made. There are definite trade-offs between having the capability to produce large numbers of products with impressive cost savings from mass production, and the flexibility to make a variety of products out of economic necessity. This is the age-old manufacturing problem, flexibility versus efficiency.

When the size of production batches steadily grows and the required time for necessary changeover diminishes, computer-automated robots become economically justified. Robots enable both requirements to be satisfied at the same time. Automation satisfies the batch-size requirement, due to the processing ability of the robot. The desired flexibility to make changeovers quickly is satisfied by the robot's reprogrammability. The robots employed in metalworking have proved many times that they can meet the two requirements efficiently. Figure 10.1 depicts graphically the flexibility versus efficiency trade-offs for different production environments, using varying levels of technology. Figure 10.1 clearly shows that when automated robots are used to produce a variety and specific quantity of different parts, the quantity of parts produced as well as the number of different parts processed is high when compared to manual production methods. It is easy to see the economic cost savings afforded by the technology. The time and money that is saved from flexibility

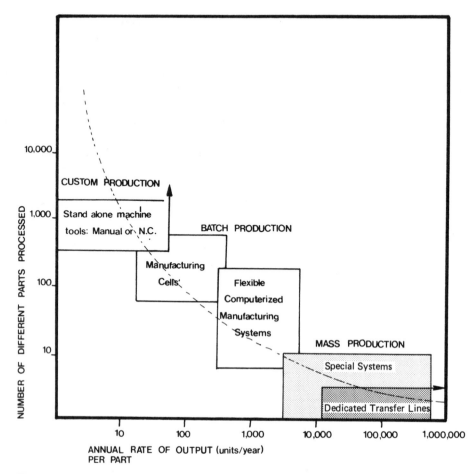

Figure 10.1 Flexibility versus efficiency trade-offs in manufacturing

alone has, in some companies, totally justified the investment in the automated robot. In addition, the competitive edge provided by the number of products and quantity potential cannot be underestimated. Companies that compete with users of automated robots have little alternative but to acquire and use the same technological edge. Those who don't will suffer economically.

Cost Versus Batch Size

In manufacturing, the relationship between the number of products that are made and the individual cost per product is affected by robotic automation, just as with traditional manufacturing methods. There are three common types of processing runs performed in industry: piece production, batch production, and mass production. Piece production involves making custom or one-of-a-kind products. When hundreds or thousands of products are manufactured, this is known as batch production. Naturally, the size of the batch is not limited. Mass production is usually applied to production runs in the hundreds of thousands or even millions. Each production run has its own characteristics as follows:

Piece Production

Very labor intensive, containing minimum application potential for computer-automated methods leading to cost savings. The fixed capital requirements are relatively stable, but the labor cost of producing each unit is unusually high. The annualized capital cost is almost negligible when compared to the annualized labor cost. The average cost per unit remains relatively constant for each unit produced, due to the uniqueness of the products.

Batch Production

Usually features semiautomated production combining automated equipment with some labor. Batch production is the middle ground where unit cost is not dominated by labor cost, fixed capital cost, and associated variable cost. Unit cost remains stable in low to mid size batches but decreases in higher batch production runs.

Mass Production

The most automated production area. It features a high capital investment in automated equipment but has the lowest associated labor costs. Annualized capital equipment cost is high. The cost to produce a unit is high when only small volumes are produced (like batch processing) but drops quickly when producing large volumes. It is the most cost-effective only when large quantities are processed. As unit throughput rates increase, so does machine utilization.

The three types of production runs contain a common characteristic: unit cost decreases as the output of units increases. The lowest unit costs are attained by producing the most units. Naturally, automation provides the capability to produce the most units, especially in mass production and high-volume batch production runs. Although the capital investment is higher, when the computer-controlled machinery is used over time, the savings provided by the lower unit cost cannot be matched by traditional labor-intensive manufacturing methods. Robotic automation is capable of producing the required throughput efficiency due to the robot's capacity for working continuously.

Labor Costs

The implementation of computer-automated manufacturing systems that include robots will ultimately displace a significant number of unskilled laborers and machine operators. Already, thousands of workers have been displaced in the United States, and tens of thousands in Japan. The potential for more displacement increases as the technology improves and more companies invest in it. As we approach the year 2000, several million workers will have been replaced by robots. As a result, there will be a decrease in direct labor input from human workers, and the unit output per manufacturer will increase. This is due to the working capacity of the robotic systems themselves. In order to estimate the true impact that robots will have in replacing factory workers and, in turn, the impact upon direct labor costs, some assumptions can be derived from possible robot uses in current industry. Typically, the workers who are being displaced today hold jobs in welding, spray painting, and foundry work, because robots are already used for these tasks.

Robots in themselves will not impact tremendous cost savings in money that is spent on the direct labor force for a long time. However, they do provide some benefits in reducing variable labor costs associated in manufacturing. Production labor cost specified in relative terms of product output often understates the actual production costs incurred when using manual or semiautomated work methods. This is because the purchase of raw materials accounts for the largest proportion of the actual product cost, as opposed to the actual costs of direct labor. Labor rates are factored on an annual basis for total parts processed and can fluctuate greatly, depending upon the size of the workforce that is needed by a manufacturer. The cost of materials, however, normally does not show this fluctuation.

As a result, the actual labor cost for producing a part or component is derived from the accumulated sum of all direct labor costs associated in the processing plus the extra labor costs that are factored into the prices of the purchased raw materials. This is due to the fact that many manufacturers buy raw materials that have already undergone a process that required direct labor. For example, an automobile manufacturer who buys steel, pays for the steel itself plus for the associated labor cost of the workers who made the steel. This labor cost is always factored in to the price of the materials.

Computer automation can provide some cost savings in this area. Reductions in direct labor costs attributed to robotic usage can be estimated by calculating the number of workers who will be displaced by robots. In factories where there is a low number of workers displaced, and with these workers coming mostly from the fabrication sectors (this sector employs skilled and unskilled labor), the estimated direct labor savings averages 2 percent in respect to the average total direct labor cost. In industries where a larger number of robots is implemented, replacing a medium number of workers, the assembly sector is affected along with the fabrication sector. Here, overall labor savings are estimated to be approximately 7 percent. The final example, where a high number of robots is used to replace a high number of workers, the displacement affects workers in fabrication, assembly, inspection, and some management workers who are employed in the materials-handling sector. In the high replacement environment, labor savings are estimated to be approximately 13 percent across the board. Of course, the replacement of even more high-paying management jobs could increase the figure significantly. Table 10.1 shows a typical breakdown of the three areas where robots replace workers and the associated impact on labor costs.

Table 10.1 Potential impact of robotics on labor costs

Industry Name	Labor Cost/ Output (%)	Reduction in Total Cost for Replacement Scenario (%)		
		Low	Medium	High
Radio, TV communication equipment	16.5	1.5	5.4	11.8
Guided missiles, space vehicles	14.3	1.5	4.8	9.5
Electronic computing equipment	8.6	1.2	3.3	6.1
Aircraft and parts	15.4	1.7	5.1	10.3
Aircraft engines and parts	19.2	2.1	6.4	12.8
Industrial patterns	42.7	5.9	16.4	30.1
Machine tool accessories	24.5	3.4	9.4	17.3
Machine tools, metalcutting	22.0	3.0	8.5	15.6
Nonferrous forgings	17.2	2.5	6.7	11.7
Residential lighting fixtures	15.9	1.4	5.2	11.4
Pumps and pumping equipment	15.7	2.2	6.0	11.1
Valves and pipe fittings	18.9	2.7	7.3	12.9
Power transmission equipment	23.2	3.2	8.9	16.4
Ball and roller bearings	27.9	3.8	10.7	19.7
Commercial laundry equipment	18.5	2.5	7.1	13.0
Blowers and fans	17.3	2.4	6.6	12.2
Travel trailers	14.2	1.5	4.7	9.5
Screw machine products	25.4	3.7	9.9	17.4
Heating equipment, not electrical	15.3	2.2	5.9	10.5
Construction equipment	16.8	2.3	6.5	11.9
Farm machinery and equipment	16.0	2.2	6.2	11.3
Truck and bus bodies	17.8	1.9	5.9	11.9
Railroad equipment	18.3	2.0	6.1	12.2
Motor vehicles, car bodies	9.8	1.1	3.3	6.5
Motor vehicle parts, accessories	22.2	2.4	7.4	14.9
Noncurrent carrying wiring devices	17.6	1.6	5.7	12.6
Metal cans	13.1	1.9	5.1	8.9
Fabricated structural metal products	17.6	2.5	6.8	12.1
Auto stampings	25.6	3.7	9.9	17.5

Implementing robots simply to achieve labor savings would not provide significant cost savings when compared to traditional manufacturing methods. Even though direct labor cost savings are not enormous when using robotic technology, the other benefits that are achieved (flexibility, capacity, quality) combine to offer substantial savings when many cost areas are considered as a whole. This may change in the future as robots with wider applicability are developed and implemented.

Capacity

The capacity to produce is where robotics and computer-aided manufacturing can realize the most impact on industry. These issues take into consideration machine and equipment usage in the factory environment. In industry today, it is fairly typical to have more machines available than operators to use them. The fact that machines must wait, sitting idle, until a human can employ them is a tremendous waste of capital. It is further compounded when you consider that the human operator is usually available for only one or two shifts, or 16 hours out of the possible 24 hours the machines could be used. As was mentioned earlier, money spent on capital equipment that sits idle is the same as paying the workers to sleep every night. When averaging the amount of time that equipment is available for work versus the actual time it is being used, the machines in our factories sit idle almost 70 percent of the total time available. Obviously, this is a tremendous waste of money and resource possibilities.

Capital equipment has the potential for use 24 hours per day, 365 days per year, or 8760 total possible work hours per year. Human operators have, on average, the potential for work (without considering overtime) of 8 hours per day, 263 days per year (because of time off for weekends, holidays, or other shutdowns) achieving a possible work potential of 2104 hours per year. Considering normal 8-hour shifts for the workers, and with a single worker per machine (this is not the case in today's industry), after the humans complete their work potential, the machines have another 6656 hours of work potential remaining. This translates into 233 total days of unproductive time that equipment sits idle. Table 10.2 contains a breakdown of equipment utilization averages in the U.S. metalworking sector for 1977. The statistics for 1977 are valid even today, because a larger number of workers was employed, and fewer automated processes were used.

The capital equipment in our factories sits idle most of the time. This is the area in which robots and computer automation stand to produce the most tangible economic benefits to indus-

Table 10.2 Average machine use in the metalworking industries in 1977

Major Group	Metal-Cutting Tools Manual (%)	Metal-Cutting Tools NCa(%)	Metal-Forming Tools (%)	Joining (Welding) (%)
34	12.4	18.4	15.8	15.1
35	12.7	20.9	7.4	17.8
36	9.9	29.3	13.0	8.2
37	16.5	14.8	17.5	40.0
Average	12.9	20.0	13.6	21.3

Sources: Machine-tool hours derived from American Machinist (1978). Labor hours derived from Bureau of Labor Statistics (1980).

aNumerically controlled (NC). Assumptions: Machine utilization is defined here as follows:

$$\text{utilization} = \frac{\text{total operator hours available}}{\text{total machine hours available}}$$

with

total operator hours available = (number of operators)*(average hours worked per operator per year)

$$\frac{\text{average hours worked}}{\text{per operator per year}} = \frac{\text{total hours worked by production workers}}{\text{total production workers}}$$

total machine hours available = (number of machines) × (8760 hr/year)

8760 hr/year = (24 hr/day)*(365 days per year)

try. Automation has the potential to increase the utilization of capital equipment significantly, thus increasing the capacity for production simply by keeping the equipment operating and producing goods. When more goods are produced, the unit cost decreases, and significant economic benefits are achieved. The potential for increasing production time to take advantage of the lost hours in equipment use is shown in Table 10.3. Increasing the utilization of machines results in more products being produced. The Japanese are already realizing these increases in several of their highly automated factories by using robot labor for the second and third shifts.

The tremendous productivity gains are realized only when an entire factory is retrofitted with computer-automated equipment. Implementing single robots in isolated workcells produces productivity increases, but they will be minor when productivity levels for the other non-automated sectors are included. To achieve maximum productivity potential and associated cost benefits, the implementation of automated systems must occur throughout the factory, in controlled automated production systems. When this occurs, there are significant benefits afforded that isolated robotic applications cannot hope to achieve. When

Table 10.3 Potential percentage increases in output from using lost time

Type of Plant	From Utilizing Days Plant is Closed	From Utilizing Nonscheduled Production Time	Total Percentage Increase in Output
High-volume	28	3	31
Mid-volume	83	115	198
Low-volume (one-shift operation)	148	187	335
Low-volume (two-shift operation)	74	43	117

robots are used with other computer-automated equipment (as in a flexible manufacturing system), production increases of 40 percent to 50 percent are typical and can be expected in most industries.

The cost to implement factory automation often runs into the hundreds of millions of dollars. This prohibits most American manufacturers from retrofitting the entire factory at one time. As automated equipment is installed, the productivity gains cited are achieved more slowly. Obviously, larger gains are realized by retrofitting the entire factory. A study conducted by the Ford Motor Company suggests that where entire factories are retrofitted with the most advanced automated technology—automated machine loading/unloading, automatic tool changing, computer diagnostic sensing devices, computer component reliability testing, unmanned robot processing—much greater productivity increases result. This study indicates that if all of these automated technologies were implemented, production capability increases almost 90 percent without increasing machining speeds. If machining speeds are increased, the potential for increase jumps to over 200 percent. These estimates include no increases in days or hours worked.

It is easy to see the significant impact that computer automation and robots can provide in increasing the utilization of capital equipment. The capacity for work is tremendously increased when robots are combined with other automated technologies. However, manufacturers should make sure that if these techniques are implemented and they begin to realize the increased benefits that the resulting increase in produced parts does not create a white elephant. Although the capacity to produce may increase, it does not necessarily mean that the markets for these goods also increases. If more goods are produced to gain the benefits afforded by automation, a glut of products can quickly become a liability. When supply exceeds demand, the price of the product drops, thus defeating the economic gains that were achieved by the automation itself. The factories of today are nowhere near facing this problem; however, the factory of 1995 could very well see this catastrophic "Catch-22" occur.

Capital Costs

Benefits in capital costs are also attained from automation. Inventories are affected by using computer automation and robots to produce more goods. These technologies facilitate the movement of pieces through the factory. In most companies involved in batch processing, of the amount of time a workpiece spends in the factory, approximately 95 percent is spent in transit or storage and only 5 percent is spent being processed or machined. When automation provides the capability to use equipment more fully, resulting in higher machine use and an increase in capacity, the amount of time required to move the piece through the factory is reduced. Robots and flexible manufacturing systems will not reduce inventory levels in a manufacturing plant, because the flow of materials through the plant will increase due to the greater capacity that automation provides. In fact, more workpieces may be necessary to keep the automated machines busy.

Keeping inventories moving through the factory provides economic benefits, because the workpiece materials are made into finished products faster. These products are sold, providing revenues in a shorter time. In other words, the faster the workpiece inventory can be converted into a finished marketable product, the sooner it can be sold, and the sooner the manufacturer recovers the cost of the workpiece material.

The capital cost impact of automation is also very significant. In most of our factories, especially those involved in some type of batch processing, capital equipment and labor costs are shared among the products that are produced. And, as discussed previously, flexibility is obtained only when equipment is taken out of production for retooling. When the equipment is being retooled, products cannot be manufactured. Since the equipment is shared among products, the retooling often affects not just one product, but several. As a result, flexibility occurs today at the economic expense of several different products that a manufacturer produces. Using computer automation and robots, especially in a flexible manufacturing system arrangement, enables the machines to be retooled through reprogramming more quickly than by using traditional methods. The equipment is back into production much sooner, again being shared, but in an increased capacity—producing larger quantities of different products. This increases the economic benefits, because several products are contributing instead of just one. The lead time required for retooling is also cut for each of the different products, which, in turn, keeps inventories of workpiece materials moving at a steady rate.

Other Issues

The economic issues that have been discussed here deal with equipment utilization, direct labor, production capacity, capital cost, and productivity. Each of these areas contributes significantly to the economic well-being of a manufacturing facility. There are other factors that automation touches that also contribute. These are the areas of quality and reliability. One of the primary reasons American industry has suffered is because the goods that we produce are often not of the same high quality as goods produced by foreign competition using automated manufacturing techniques. Computer automation is known for speed and accuracy. The use of robots in spray painting and welding have already proven that quality can be achieved, because of the machine's ability to perform the same process repeatedly with the same precision and accuracy. Uniform precision enhances quality. Higher quality equals increased reliability.

As we begin to use more automated systems to produce goods, the precision of these goods will improve along with the quality. For the purchasers of these goods—consumers, government, or industry—improved quality of domestically produced items means that more American products will be bought. This will occur here and abroad. When more American products are purchased, the resulting dollars are returned to our economy, which facilitates growth. Over the long run, the revitalization will cause renewed growth in the manufacturing sector, which will improve our national economy as a whole, as well as reduce (perhaps halt) the growing trade deficit. This example is greatly simplified but it represents what could happen if we use the benefits that are attainable from automated manufacturing. After all, the Japanese have used it against us successfully.

The Future

The year 2000 will be here in less than 15 years. It seems like a long way off, but it was only 15 years ago that Richard Nixon was president, computers were mysterious and aroused curosity, and foreign cars raised eyebrows. If technology continues to advance as it has during the past 15 years, the capabilities and uses for robots will grow significantly. As a result, so will the number of robots. Just as the computer itself has become a commonplace item, advertised in magazines and on televison, robots probably will too. Our reluctance to use robotic technology will have passed, and we will find these mechanical workers in more areas than are possible, or even imaginable, with today's level of sophistication.

Robots Themselves

Robots are a part of the information age. They use the power of the computer to function. As the computer becomes more advanced, so too will robots. We will find them in environments far removed from the factory floor where they were first introduced over 25 years ago. However, it is the experiences we gain from their use in factories that will be used to further expand their role in other environments and applications. We will see robots being used for nuclear maintenance and cleanup, waste disposal/garbage disposal, security, medical applications, space exploration, undersea exploration, warfare, and many other areas that are dangerous or distasteful to humans.

Just as robots evolved from their early pick-and-place abilities, the robots of the 1990s will see broadened capabilities as a result of increased computer control. Robot manufacturers predict that by 1990 the U.S. will be producing over 17,000 robots per year. There will be over 80,000 robots working in industry, a far cry from the 15,000 we currently use. Robot implementation is expected to increase approximately 35 percent per year through the end of the decade.

Robots will evolve in function so that they can perform a wider variety of tasks. They will become much more versatile, able to handle complex tasks more easily. Where today's robots have limitations due to restrictions of their end effectors, the robots of tomorrow will have greater tactile sensing control. Gripping functions will expand, making robot hands similar to human hands—in looks as well as function. Improved tactile sensing will include orientation, recognition, and wider ranges of physical interaction. End effectors will attain many of the sensing abilities we take for granted as humans. This will enable them to be used for many more tasks that require a delicate touch, something that the end effectors of today cannot accomplish.

We will see the introduction of end effectors with proportional mobility, provided by enhancements in servocontrol. This will give robot hands more freedom of motion and manipulative control than seen today. The development of modular end effectors is also expected to become a reality. Modularity will enable robots to use a variety of different hands, increasing their capacity to perform different functions. End effectors will simply plug in and out, being changed in seconds, without requiring any rewiring or other modifications that would affect production or interrupt tasks. Naturally, flexibility will increase with the added capabilities that multiple end effectors create. This will allow a single robot to be used for a variety of different applications merely by changing the robot's hands and providing it with different functional instructions from a computer.

Robots will branch out beyond the industrial arm-type manipulator and end effector of today to become more versatile because they will be made from lighter weight materials. The lighter weight will enable robots to achieve even higher degrees of repeatability with far more accuracy and precision. This, combined with improved tactile sensing, will enable robots to aid in assembly of fit-ups and provide for the rejection of misfitting components or parts. Higher levels of compliance will also result where robots can bring parts or components together without causing damage. Some individuals envision the day when robot hands are so humanlike and maneuverable that they can be used for such touch-sensitive jobs as semiconductor packaging and even microsurgery. More uses for laser technology are expected to enter the factory via robots. Robots will have lasers as end effectors. Controlled by computers, they will be used in welding, cutting, and fusing applications.

Robots will look different, too. They will be less bulky and more mobile. As they evolve, some will acquire more humanlike qualities (perhaps to achieve increased acceptance), while most will still look like the machines that they are. To truly look human in appearance, robots will naturally require two arms that are capable of careful coordination. New designs will enable them to handle greater loads easily and with more precision. New motor technology will provide for wider ranges of motion that are almost graceful. This will allow more intricate functions to be performed where space restrictions are a problem.

The area that stimulates the most excitement and interest is in giving robots vision capabilities. Here, they will truly personify humans. Vision could be the watershed for robotics. Advances in computer technology will provide "thinking" eyes that are able to react to environmental stimuli. Rudimentary vision will be derived from software that contains orientation data as well as recognition data, making the robot able to perceive the objects it examines. Stereo vision, like that of humans, will make robots capable of inspection and testing functions to meet a variety of needs, in and out of the factory. Vision will create near "android" types of robots when it is combined with voice capabilities. Voice recognition and computer-generated voice synthesis are available today. Robots that can see and talk are very close to the next evolutionary process, robots that can think.

Artificial Intelligence and Robots

Computers will be much more efficient in 20 years. The era of the "thinking machine" is upon us today, with remarkable advancements in artificial intelligence and expert systems. Artificial intelligence concepts, where computers learn from their experiences just like a human, will make tomorrow's robots smarter. They will be able to learn from their previous experiences and adjust themselves automatically through self-reprogramming. Smart robots, interacting with their environment will not be so much like the robots from the science fiction movies, the "take me to your leader" type, but functioning, helpful robots that are capable of performing work and thinking for themselves (through the computer).

Artificial intelligence (AI), once mastered, will produce the next generation in robots. Computers with AI capabilities will be able to more effectively control and manipulate a wider variety of end effectors for different purposes that require learning. There will be a tremendous increase in flexibility offered by the technology. Robotics may very well be the first successful major commercial use of artificial intelligence.

Robots in the Factory

Manufacturing will continue to employ robots in jobs where humans are exposed to dangers, hazards, and other risks. However, robots will also expand into many other factory sectors where they have previously been absent. Existing robotic technologies will be exploited to the fullest in the coming years. Industry will not have to wait until a robot application is perfected before it can be used for productive work. We will see new

Robot Generations

First Generation—These robots were primitive, operating by fixed-stops control with a limited number of movements. They were used exclusively for pick-and-place tasks in engineering industries and are still used in industry. The Unimation 2000 is an example of a first generation robot.

Second Generation—In this generation, the robots were programmed using a button box. The task sequence that the robot was to perform was recorded on magnetic tape. When the tape is played back, using the button box, the robot "learns" the operation that it is to perform. As the tape cycles, the robot repeats the functions until the tape is stopped. Most of the second generation robots are used in spray painting and spot welding in the automobile industry. Examples of this type of robot are Cincinnati Millacron's T³ ("The Tomorrow Tool"), robots made by Asea, paint-spraying units by Trallfa, Germany's KUKA robots, and Japan's Hitachi robots.

Third Generation—These were the first truly computer-programmable robots. They received their operating instructions directly from the communication links with the controlling computer. They have been commercially available since 1981. Third-generation robots have limited sensing capabilities and are programmed with impenetrable coordinate systems, rather than in terms of objects. Examples of these robots are the Unimation PUMA robot, which is programmed using VAL (a robot programming language), robots by Automatix programmed in RAIL (another programming language), and IBM's 7535 and 7565 robots.

Fourth Generation—This is the next generation of robots. These machines are under development by robotic vendors today. They will feature combined end effectors (arms, legs, jigs, and feeders) that work together in a coordinated effort. Tactile capabilities will improve significantly, giving robots the ability for soft touch. Artificial intelligence will be heavily utilized to solve the problems associated in robot learning and training.

levels of sophistication in robots that are developed to solve robotic problems we face today. The industrial robot of tomorrow will come from the manufacturing proving grounds of today. Where more repeatability is desired, more repeatability will be designed; where increased flexibility is needed, flexibility will be enhanced. Today's application problems will be tomorrow's application realities.

Computers and robots will be in use throughout the factory, not isolated into islands, as is the case today. The different sectors will be integrated electronically, so that information can be exchanged among all sectors, making the factory operate more efficiently. We will begin to see computer-integrated manufacturing on a broad level. As the price of computers and robots decreases, manufacturers will be able to use even more of these machines to work in and control all areas of a factory. Networks will link factory equipment and robots together, providing control over factory production. Only a computer can work fast enough to process and track the flow of inventory, from raw materials to finished goods, through the manufacturing process while monitoring the activities of the robots and automated machine tools at the same time.

The manufacturing process will change to accommodate more computer-aided design input that ultimately speeds up the computer-aided manufacturing cycle. Once a part is designed, the computer will fully test it in a computer simulation. After the design process is complete, the computer coordinates from the design phases can be transferred automatically to the machining centers. Here, total computer control will process parts and keep the production cycles moving smoothly. Enhanced equipment will require little operator intervention, as tools will be redesigned to make automatic tool changing easy.

Many of our factories will develop into mega-assembly systems. These factories will feature integrated flexible systems, including multistation, multiproduct, progressive assembly stations that use 10 or more robots and automated tools. These assembly stations will be linked electronically, so that each can adjust automatically to situations created by neighboring stations that share applications. The Japanese already use early mega-assembly systems; however, they lack the total computer integration that future assembly systems will contain. Some of the larger mega-assembly factories will contain 200 stations, where at least one-third will feature robots. These factories will be responsible for mass production and high-volume batch processing. They will have extreme flexibility in order to change and adapt to produce new products. Human workers will play an important role also, but not as traditional workers. In the mega-assembly factory, humans will take on the role of the caretaker of the machinery (of which there will be plenty). They will provide computer-programming support, computer maintenance, machine maintenance, and other service-related tasks.

Computers will greatly enhance the capability to store libraries of information regarding products that are manufactured, much as a computer-aided design system does. Here, figures are designed electronically, plotted, and then archived onto some type of storage device, either magnetic tape or disk, where they are saved until they are needed again. This saves countless hours of work if the same design is needed in the future. The figures can be called up and used whenever they are needed. Changes to existing figures can be made quickly and easily. Computer programs for manufacturing production will develop in the same manner. Once a product is manufactured, the program containing the manufacturing instructions for the robots and automated tools will be saved like the CAD design. Whenever the same part is made again in the future, the program is retrieved and run, and the part can be made. If a part changes in some way, the instructions can be modified and the new part produced quickly. As factories develop more products or change existing products, the libraries of stored programs will increase. Some manufacturing experts predict that the product design process will focus on creating more individual component shapes that will use group technology and expansive design graphics data bases. This information can then be translated into machine-processing information and electronically sent to the correct factory sector.

In product design, sweeping changes will also occur. As computer-integrated manufacturing increases, product design will have to change to allow for the more automated methods. The majority of the products that are manufactured today cannot be manufactured using automated techniques, because their inherent design prohibits it. They are not designed for efficient production methods. As a result, many of these products will be totally redesigned with automation in mind. The design will take into consideration the manufacturing issues that involve integration between sectors of the factory. For example, a product that is being designed for automated fabrication will likely be designed for automated assembly operations as well. By redesigning products, many of the barriers to using computer-automated manufacturing techniques will be overcome. This will allow more products to be manufactured using automated processes.

New programming languages will be developed to make the most efficient use of computer automation. As these languages are perfected, they will offer more power and capability than the robotic software used today. Programming languages will come out of university and industry research laboratories and go to work in the factory. These new languages will promote the execution of more complex tasks by the machines. Stanford University is involved in the development of these advanced robotic languages. Some of these languages will operate in computers equipped with artificial intelligence capabilities, enabling robots to actually "learn" from their mistakes to become more productive and efficient, and even take their commands in the form of verbal instructions from a human. The concepts and implementation of artificial intelligence in robotic application research and development work occurring today is structured in making the robot a part of the entire factory. Robots will become a very important part of the entire manufacturing process in the coming years.

Computer automation will provide industry with a knowledge base of operations. This has been missing in our manufacturing facilities almost from their creation. In the factory of yesterday, different sectors had little reason to communicate and share work methods. Automation changes this. Where before, techniques were relatively isolated within sectors, automation provides for the sharing of work methods and experiences through electronic commands. Knowledge, in the form of electronic information, will make managing a manufacturing facility easier. Factories will be required to develop methods to gather and input information into the computers. This will provide new employment possibilities for workers who are displaced by the technology. In the same way as office personnel learned to use the computer, so too will factory workers and managers.

Factory automation in the future will require an entirely different form of manager. The manager of the future will possess a broad understanding of computer technology and its effects in the factory environment. Colleges and universities will redevelop their curricula so that future graduates will be schooled in automated-factory techniques. The manager of tomorrow will be a product of the information age and be required to make decisions based upon computer technology as well as manufacturing methods. These managers will be responsible for the organization and operation of complex flexible systems using larger computer data bases that run not only the robots but coordinate the entire factory. Managers today will be required to relearn control processes and think in terms of automated processes. Paper shuffling, common today for operational control, will be reduced or totally eliminated by automation.

Automation will also affect industrial management's decision-making abilities. Managers will have more data from which they can develop manufacturing strategies. Computers will provide information regarding all of the operations within the factory by monitoring the machines, processing, inventories, and schedules. At the touch of a finger, a manager can find out what is happening in the fabrication sector on the factory floor. Inventory information will be up-to-the-minute, and materials-handling programs will keep constant tabs on all materials and workpieces as they proceed through the manufacturing process.

By the year 2000, we will be well on our way toward the reality of the factory of the future. The capabilities of computers and robots will undoubtedly eclipse the technology we currently enjoy. And, as technology increases, so too will the applications for these machines. The factory of today will be a memory. Our ways of thinking about manufacturing, as well as our manufacturing methods themselves, will be totally changed. Where today robots are somewhat of a curiosity, tomorrow they may very well be in our homes, becoming just another machine whose use we take for granted.

APPENDIX A

Suggested Readings

Abodaher, David. *Iacocca*. New York: Macmillan Publishing Company, 1982.

Asimov, I. *I, Robot*. New York: Doubleday, 1950.

Ayres, R.U., and Miller, S.M. *Robotics, Applications and Social Implications*. Cambridge MA: Ballinger Publishing Company: 1983.

Berger, Phil. *The State-of-the-Art Robot Catalog*. New York: Dodd, Mead & Company, 1984.

Drucker, P.F. "The Re-Industrialization of America." *The Wall Street Journal*, June 13, 1980, p. 10.

Engelberger, J.F. *Robotics in Practice*. London: Kogan Page Ltd., 1980.

Japan Industrial Robot Association. *The Robotics Industry of Japan: Today and Tomorrow*. Englewood Cliffs, New Jersey: Prentice-Hall, 1982.

Naisbitt, John. *Megatrends: Ten New Directions for Transforming Our Lives*. New York: Warner Books Inc., 1892.

Newspaper Enterprise Association, Inc. *The World Almanac & Book of Facts 1985*. New York: Newspaper Enterprise Association, Inc., New York, 1985.

Nof, Shimon Y., *Handbook of Industrial Robotics*. Ed. New York: John Wiley & Sons Inc., 1985.

Skinner, Wickham. *Manufacturing: The Formidable Competitive Weapon*. New York: John Wiley & Sons, Inc., 1985.

Winston, Patrick H., *The AI Business: The Commercial Uses of Artificial Intelligence*. ed. New York: The MIT Press, 1984.

International Standard Organization, Selected papers 1976-1985.

Glossary

actuator: A motor or transducer that converts energy (electrical, hydraulic, or pneumatic) into power used to produce motion or power.

accuracy: The ability of a robot to position its end effector at a programmed location in space.

analog control: Control signals that are processed by directly measuring quantities (voltages, resistances, or rotations). This control can be hydraulic, electronic, or pneumatic.

android: A robot or machine that resembles a human.

anthropomorphic robot: A robot with rotary joints that can move much like a person's arm.

arm: An interconnected set of links and powered joints comprising a manipulator that supports or moves a wrist, hand, or end effector.

artificial intelligence: The ability of a machine system to perceive anticipated or unanticipated new conditions, decide what actions must be performed under the conditions, and plan the actions accordingly. The main areas of applications are expert systems and computer vision.

assembly: Also known as an assembly cell, or assembly station. A concentrated group of equipment, such as manipulators, vision modules, parts presenters, and support tables, that are dedicated to complete assembly operations at one physical location. A computer and a control terminal complete the system. In operation, one or more robot arms completes the assembly of the product while, if needed, other arms simultaneously preassemble the next product and/or prepare subassemblies.

automated inspection: The use of any one of several techniques to determine the presence or absence of features. The techniques include simple mechanical probes and vision systems.

automation: Automatically controlled operation of an apparatus, process, or system by mechanical or electronic devices that replace human observation, effort, and decision.

axis: A traveled path in space, usually referred to as a linear direction of travel in any of three dimensions. In Cartesian coordinate systems, labels of x, y, and z are commonly used to depict axis directions relative to earth. X refers to a directional plane or line parallel to earth, y refers to a directional plane of line that is parallel to earth and perpendicular to x, and z refers to a directional plane or line that is vertical to and perpendicular to the earth's surface.

bang-bang robot: A robot in which motions are controlled by driving each axis or degree of freedom against a mechanical limit stop.

base: The platform or structure to which a robot arm is attached; the end of a kinematic chain of arm links and joints opposite to that which grasps or processes external objects.

batch manufacturing: The production of parts or materials in discrete runs, or batches, interspersed with other production operations or runs of other parts or materials.

CAD/CAM: An acronym for computer-aided design and computer-aided manufacturing.

cell: A manufacturing unit consisting of two or more workstations or machines, and the material transport mechanisms and storage buffers that interconnect them.

center: A manufacturing unit consisting of two or more cells, and the material transport mechanisms and storage buffers that interconnect them.

complex joint: An assembly between two closely related rigid members enabling one member to rotate in relation to the other around a mobile axis.

computer-aided design (CAD): The use of an interactive-terminal workstation, usually with graphics capability, to automate the design of products. CAD includes functions such as drafting and fit-up of parts.

computer-aided manufacturing (CAM): Working from a product design likely to exist in a CAD data base, CAM encompasses the computer-based technologies that physically produce the product, including parts-program preparation, process planning, tool design, process analysis, and parts processing by numerically controlled machines.

computer-integrated manufacturing (CIM): The philosophy dictating that all functions within a manufacturing operation be data base–driven and that information from within any single data base be shared by other functional groups. CIM includes major functions of operations management (purchasing, inventory management, order entry), design and manufacturing engineering (CAD, NC programming, CAM), manufacturing (scheduling, fabrication, robotics, assembly, inspection, materials handling), and storage and retrieval (inventories, incoming inspection, shipping, vendor parts).

computerized numerical control (CNC): A numerical control system with a dedicated mini- or micro computer that performs the functions of data processing and control.

cycle (program): The unit of automatic work for a robot. Within a cycle, subelements called trajectories define lesser but integral elements. Each trajectory is made up of points where the robot performs an operation or passes through, depending upon the programming.

degrees of freedom: The number of independent ways the end effector can move. It is defined by the number of rotational or translational axes through which motion can be obtained. Every variable representing a degree of freedom must be specified if the physical state of the manipulator is to be completely defined.

design process (of robots): A multistep process beginning with a description of the range of tasks to be performed. Several viable alternative configurations are then determined, followed by an evaluation of the configurations with respect to the sizing of components and dynamic system performance. Based upon the appropriate technical and economic criteria, a configuration can be selected. If no configuration meets the criteria, the process may be repeated in an iterative manner until a configuration is selected.

elbow: The joint that connects the upper arm and forearm on a robot.

end effector: Also known as end-of-arm tooling or, more simply, a hand. The subsystem of an industrial robot system that links the mechanical portion of the robot (manipulator) to the part being handled or worked on, and gives the robot the ability to pick up and transfer parts and/or handle a multitude of differing tools to perform work on parts. It is commonly made up of four distince elements: a method of attachment of the hand or tool to the robot tool-mounting plate, power for actuation of tooling machines, mechanical linkages, and sensors integrated into the tooling. Examples include grippers, paint spraying nozzles, welding guns, and laser gauging devices.

expert system: A computer program, usually based on artificial intelligence techniques, that performs decision-making functions that are similar to those of a human expert and, on demand, can justify to the user its line of reasoning. Typical applications in the field of robotics are high-level robot programming, planning and control of assembly, and processing and recovery of errors.

first-generation robot system: Robots with little, if any, computer power. Their only intelligent functions consist of learning a sequence of manipulative actions, choreographed by a human operator using a teach-box. The factory world around them has to be prearranged to accommodate their actions. Necessary constraints include precise workpiece positioning, care in specifying spatial relationships with other machines, and safety for nearby humans and equipment.

fixed-stop robot: Also known as a nonservo robot or open loop robot. A robot with stop-point control but no trajectory control. Each of its axes has a fixed mechanical limit at each end of its stroke and can stop only at one or the other of these limits. See also bang-bang robot.

flexibility, operational: Multipurpose robots that are adaptable and capable of being redirected, trained, or used for new purposes. Refers to the reprogrammability of multitask capability of robots.

flexible manufacturing system (FMS): An arrangement of machine tools that is capable of standing alone, interconnected by a workpiece transport system, and controlled by a central computer. The transport subsystem, possibly including one or more robots, carries work to the machines on pallets or other interface units so that accurate registration is rapid and automatic. FMS may have a variety of parts being processed at one time.

floor-mounted robot: Also known as a pedestal robot. A robot with its base permanently or semipermanently attached to the floor or a bench. Such a robot is working at one location with a maximum limited work area and in many cases servicing only one machine. Floor-mounted robots often use a pallet pick-and-place or a conveyor feeder to feed parts to and from their location.

forearm: That portion of a jointed arm which is connected to the wrist and elbow.

gantry robot: An overhead-mounted, rectilinear robot with a minimum of three degrees of freedom and normally not exceeding six. Bench-mounted assembly robots that have a gantry design are not included in this definition. A gantry robot can move along its x- and y-axes traveling over relatively greater distances than a pedestal-mounted robot and at high traverse speeds while still providing a high degree of accuracy for positioning. Features of a gantry robot include large work envelopes, heavy payloads, mobile overhead mounting, and the capability and flexibility to operate over the work area of several pedestal-mounted robots.

gripper: The grasping hand of the robot, which manipulates objects and tools to fulfill a given task.

group technology: A technique for grouping parts to gain design and operational advantages. For example, in robotics, group technology is used to ensure that different parts are of the same parts family when planning part processing for a workcell, or to design widely usable fixtures for parts families. Parts grouping may be based on geometric shapes, operation processes, or both.

hand: A fingered gripper sometimes distinguished from a regular gripper by having more than three fingers and more dexterous finger motions resembling the human hand.

hard automation: Also known as fixed automation, or hard tooling. A nonprogrammable, fixed tooling that is designed and dedicated for specific operations that are not easily changeable. It may be configured mechanically and is cost-effective for a high production rate.

inspection (robotic): Robot manipulation and sensory feedback to check the compliance of a part or assembly with specifications. In such applications, robots are used in conjunction with sensors, such as a television camera, laser, or ultrasonic detector, to check parts locations, identify defects, or recognize parts for sorting. Application examples include inspection of printed circuit boards, valve cover assemblies for automotive engines, sorting of metal castings, and inspection of the dimensional accuracy of openings in automotive bodies.

islands of automation: An approach used to introduce factory automation technology into manufacturing by selective application of automation. Examples include numerically controlled machine tools; robots for assembly, inspection, painting, or welding; automated assembly equipment; and flexible machining systems. Islands of automation should not be viewed as ends in themselves but as a means of forming integrated factory systems. They may range in size from an individual machine or workstation to entire departments.

job shop: A discrete parts-manufacturing facility characterized by a mix of products of relatively low-volume production in batch lots.

level of automation: The degree to which a process has been made automatic. Relevant to the level of automation are questions of automatic failure recovery, the variety of situations that will be automatically handled, and the conditions under which manual intervention or action by human beings is required.

limit switch: An electrical switch positioned to be switched when a motion limit occurs, thereby deactivating the actuator that causes the motion.

machine loading/unloading (robotic): The use of the robot's manipulative and transport capabilities in ways generally more sophisticated than simple materials-handling. Robots can be used to grasp a workpiece from a conveyor belt, lift it to a machine, orient it correctly, and then insert it or place it on the machine. After processing, the robot unloads the workpiece and transfers it to another machine or conveyor.

machining center: A numerically controlled metal-cutting machine tool that uses tools, such as drills or milling cutters, equipped with an automatic tool-changing device to exchange those tools for different and/or fresh ones. In some machining centers, programmable pallets for parts fixturing are also available.

manipulator: A mechanism, usually consisting of a series of segments, or links, jointed or sliding relative to one another, for grasping and moving objects, usually in several degrees of freedom. Is is remotely controlled by a human (manual manipulator) or a computer (programmable manipulator). A manipulator refers mainly to the mechanical aspect of a robot.

materials-handling (robotic): The use of the robot's basic capability to transport objects. Typically, motion takes place in two or three dimensions, with the robot mounted stationary on the floor, on slides or rails that enable it to move from one workstation to another, or overhead. Robots used in purely materials-handling operations are typically nonservo or pick-and-place robots. Some application examples include transferring parts from one conveyor to another, transferring parts form a processing line to a conveyor, palletizing parts, and loading bins and fixtures for subsequent processing.

microprocessor: The basic element of a central processing unit that is constructed as a single integrated circuit.

mounting plate: The means of attaching end-of-arm tooling to an industrial robot. It is located at the end of the last axis of motion on the robot. The mounting plate is sometimes used with an adapter plate to enable the use of a wide range of tools and tool power sources.

numerical control: A method for the control of machine tool systems. A program containing all the information, in symbolic numerical form, needed for processing a workpiece is stored on a medium such as paper or magnetic tape. The information is read into a computer controller, which translates the program instructions to machine operations on the workpiece. See also computerized numerical control.

off-line programming: Developing robot programs partially or completely without requiring the use of the robot itself. The program is loaded into the robot's controller for subsequent automatic action of the manipulator. An off-line programming system typically has three main components: geometric modeler, robot modeler, and programming method.

operating system: Software that controls the execution of computer programs; may provide scheduling, allocation, debugging, data management, and other functions.

orientation: Also known as positioning. The consistent movement or manipulation of an object into a controlled position and attitude in space.

palletizing/depalletizing: A term used for loading/unloading a carton, container, or pallet with parts in organized rows and possibly in multiple layers.

payload: The maximum weight that a robot can handle satisfactorily during its normal operations and extensions.

peripheral equipment: The equipment used in conjunction with the robot for a complete robotic system. This includes grippers, conveyors, parts positioners, and parts or materials feeders that are needed with the robot.

pick-and-place: A grasp-and-release task, usually involving a positioning task.

pick-and-place robot: Also know as bang-bang robot. A simple robot, often with only two moves or three degrees of freedom, that transfers items from a source to a destination via point-to-point moves.

pitch: Also known as bend. The angular rotation of a moving body about an axis that is perpendicular to its direction of motion and in the same plane as its top side.

pixel: Also known as photoelement, or photosite. A digital picture or sensor element, pixel is short for picture cell.

programmable automation: Automation for discrete parts manufacturing characterized by the features of flexibility to perform different actions for a variety of tasks, ease of programming to execute a desired task, and artificial intelligence to perceive new conditions, decide what actions must be performed under those conditions, and plan the actions accordingly.

programmable manipulator: A mechanism that is capable of manipulating objects by executing a program stored in its control computer (as opposed to a mechanical manipulator, which is controlled by a human).

programming (robot): The act of providing the control instructions required for a robot to perform its intended task.

repeatability: The envelope of variance of the robot tool point position for repeated cycles under the same conditions. It is obtained from the deviation between the positions and orientations reached at the end of several similar cycles. Contrast with accuracy.

robot: A reprogrammable, multifunctional manipulator designed to move material, parts, tools, or specialized devices through variable programmed motions for the performance of a variety of tasks.

robotics: The science of designing, building, and applying robots.

robot systems: A robot system includes the robot(s) (hardware and software) consisting of the manipulator, power supply, and controller; the end effector(s); any equipment, devices, and sensors required for the robot to perform its task; and any communications interface that is operating and monitoring the robot, equipment, and sensors. (This definition excludes the rest of the operating system hardware and software.)

roll: Also known as twist. The rotational displacement of a joint around the principal axis of its motion, particularly at the wrist.

second-generation robot systems: A robot with a computer processor added to the robotic controller. This addition makes it possible to perform, in real time, the calculations required to control the motions of each degree of freedom in a cooperative manner to effect smooth motions of the end effector along predetermined paths. It also becomes possible to integrate simple sensors, such as force, torque, and proximity, into the robot system, providing some degree of adaptability to the robot's environment.

sensor: A device, such as a transducer, that detects a physical phenomenon and relays information to a control device.

sensory control: The control of a robot based on sensor readings.

sequencer: A controller that operates an application through a fixed sequence of events.

servocontrolled robot: A robot driven by servomechanisms; that is, motors or actuators whose driving signal is a function of the difference between a commanded position and/or rate and measured actual position and/or rate. Such a robot is capable of stopping at or moving through a practically unlimited number of points in executing a programmed trajectory.

servomechanism: An automatic control mechanism consisting of a motor or actuator driven by a signal that is a function of the difference between a commanded position and/or rate and measured actual position and/or rate.

shoulder: The manipulator arm linkage joint that is attached to the base.

slew rate: The maximum velocity at which a manipulator joint can move; a rate imposed by saturation in the servoloop controlling the joint.

spot-welding robot: A robot used for spot welding and consisting of three main parts: a mechanical structure comprising the body, arm, and wrist; a welding tool; and a control unit. The mechanical structure serves to position the welding tool at any point within the working volume and orient the tool in any given direction so that it can perform the appropriate task.

spraying (robotic): Robot manipulation of a spray gun to apply some material, such as paint, stain, or plastic powder, to either a stationary or moving part. These coatings are applied to a wide variety of parts, including automotive body panels, appliances, and furniture. Other uses include the application of resin and chopped glass fiber to molds for producing glass-reinforced plastic parts and spraying epoxy resin between layers of graphite broad goods in the production of advanced composites.

stepping motor: A bidirectional, permanent-magnet motor that turns through one angular increment for each pulse applied to it.

stereo imagin: The use of two or more cameras to pinpoint the location of an object point in a three-dimensional space.

stop (mechanical): A mechanical constraint or limit on some motion. It can be set to stop the motion at a desired point.

tactile sensing: The detection by a robot through contact of touch, force, pattern slip, and movement. Tactile sensing allows for the determination of local shape, orientation, and feedback forces of a grasped workpiece.

teach programming: Also known as teaching. A method of entering a desired control program into the robot controller. The robot is manually moved by a teach pendant (a hand-held programming device) or led through a desired sequence of motions by an operator. The movement information as well as other necessary data is recorded by the robot controller as the robot is guided through the desired path.

third-generation robot system: A robot system characterized by the incorporation of multiple computer processors, each operating asynchronously to perform specific functions. A typical third-generation robot system includes a separate low-level processor for each degree of freedom and a master computer supervising and coordinating these processors, as well as providing higher-level functions.

tilt: The orientation of a view, as with a video camera, in elevation.

tool changing (robotic): An alternative to dedicated, automatic tool changers that may be attractive because of increased flexibility and a relatively lower cost. A robot equipped with special grippers can handle a large variety of tools that can be shared quickly and economically by several machines.

tracking: A continuous position-control response to continuously changing input requirements.

translation: A movement such that all axes remain parallel to what they were (i.e., without rotation).

transport (robotic): The acquisition, movement through space, and release of an object by a robot. Simple materials-handling tasks requiring one- or two-dimensional movements are often performed by nonservo robots. More complicated operations, such as machine loading/unloading, palletizing, parts sorting, and packaging, are typically performed by servocontrolled, point-to-point robots.

upper arm: The portion of a jointed arm that is connected to the shoulder.

vision system: A system interfaced with a robot that locates a part, identifies it, directs the gripper to a suitable grasping position, picks up the part, and brings the part to the work area. A coordinate transformation between the camera and the robot must be carried out to enable proper operation of the system. Vision systems are also used for manufacturing inspection tasks.

welding (robotic): Robot manipulation of a welding tool for spot or arc welding. Robots are used in welding applications to reduce costs by eliminating human labor, improve product quality through better welds, and, particularly in arc welding, to minimize human exposure to harsh environments. Spot welding of automotive bodies, normally performed by point-to-point servorobots, is currently the largest single application for robots.

work envelope: Also known as the robot-operating envelope. The set of points representing the maximum extent or reach of the robot tool in all directions.

workspace: The envelope reached by the center of the interface between the wrist and the tool, using all available axis motions.

wrist: A set of joints, usually rotational, between the arm and the hand or end effector, which allow the hand or end effector to be oriented relative to the workpiece.

yaw: The angular displacement of a moving joint about an axis which is perpendicular to the line of motion and the top side of the body.

APPENDIX C

Reference Tables

Table C.1 World Robot Population, End Of 1982.

	Number	Percent
Japan	18,000	51
United States	6,200	18
Western Europe		
West Germany	2,800	8
Sweden	1,600	5
United Kingdom	800	2
France	700	2
Italy	500	1
Norway	400	1
Other	400	1
U.S.S.R.	3,000	9
Eastern Europe	600	2
Total	35,000	100

Table C.2 Robot density by country

Country	Robots per 10,000 Employed in Manufacturing			
	1974	1978	1980	1981
Sweden	1.3	13.2	18.7	29.9
Japan	1.9	4.2	8.3	13.0
W. Germany	0.4	0.9	2.3	4.9
United States	0.8	2.1	3.1	4.0
France	0.1	0.2	1.1	1.9
United Kingdom	0.1	0.2	0.6	1.2

Table C.3 Cumulative installed robots in U.S. industry

	1980	1981	1982	1983	1985 f	1990 f
Cumulative Installed (year-end)	3100	4500	6200	8200	14400	30000

(f = forecasted)

Table C.4 Robot market share worldwide

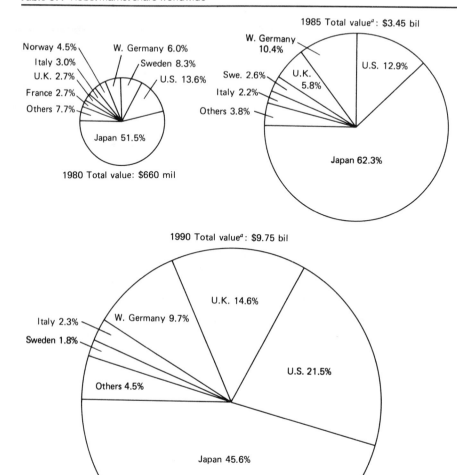

Norway 4.5%
Italy 3.0%
U.K. 2.7%
France 2.7%
Others 7.7%
W. Germany 6.0%
Sweden 8.3%
U.S. 13.6%
Japan 51.5%

1980 Total value: $660 mil

1985 Total value^a: $3.45 bil

W. Germany 10.4%
Swe. 2.6%
U.K. 5.8%
Italy 2.2%
Others 3.8%
U.S. 12.9%
Japan 62.3%

1990 Total value^a: $9.75 bil

U.K. 14.6%
W. Germany 9.7%
Italy 2.3%
Sweden 1.8%
Others 4.5%
U.S. 21.5%
Japan 45.6%

^{a}Estimated

Table C.5 Installed operating industrial robots

| | By Application | | |
| | Japan | United States | |
	(1982)	(1982)	(1990 f)
Welding	25%	35%	23%
Materials-Handling	20%	20%	12%
Assembly	20%	2%	12%
Machine Loading	8%	15%	19%

(continued)

Table C.5 Continued

	By Application		
	Japan	United States	
	(1982)	(1982)	(1990 f)
Painting	3%	10%	6%
Foundry	2%	15%	11%
Other	22%	3%	17%
Totals	100%	100%	100%

(f = forecast)

		By Industry
SIC*	Industry	United States
33	Primary Metals	18.0%
34	Fabricated Metals	15.0%
35	Machinery, Nonelectrical	11.0%
36	Electrical Machinery	3.0%
36	Electronics	3.0%
37	Automotive	43.0%
37	Aerospace	0.5%
	All Others	6.5%
	Total	100.0%

(* Standard Industrial Classification)

Robotic Organizations and Manufacturers

Robotic Associations

Belgium
Belgian Industrial Robot Association (BIRA)
c/o Chef de Service/DEI
Fabrique Nationale—Brugge
Ten Briele 2
B–8200 Brugge
Belgium

Denmark
Danish Robot Association
c/o Technological Institute
Division of Industrial Automation
Gregersensvej, 2630 Taastrup
Denmark

Finland
Robotics Society
Oy Nokia Ab Robotics
Pursimiehenkatu 29-31
SF–00150 Helsinki 15,
Finland

France
Association Francaise de Robotique Industrieele
(AFRI)
89 Rue Falguerie
75015 Paris
France

Italy
Societa Italiana Robotica Industriale (SIRI)
Instituto di Electrotechnica ed Electronics
Politechnico di Milano
Piazza Leonardo da Vinci 32
21022 Milano
Italy

Japan
Japan Industrial Robot Association (JIRA)
Kikai Shinko Kaikan Building
3–5–8 Shiba-kown
Minato-ku
Tokyo 105
Japan

Netherlands
Contactgroep Insutriele Robots (CIR)
Landgbergstratt 3
2628 CE Delft
Netherlands

Singapore
Singapore Robotic Association
5, Portsdown Road
Off Ayer Rajah road
Singapore 0513
Republic of Singapore

Spain
Comite Espanol de robots Industriales
c/o Instituto do Automatica Industrial (C.S.I.C.)
CRA Valencia KM.22800—la Poveda
Arganda de Ray (Madrid)
Spain

Sweden
Swedish Industrial Robot Association
Box 5506
Storgatan 19
S–114 85 Stockholm
Sweden

United Kingdom
British Robot Association (BRA)
35–39 High Street
Kempston Bedford MK42 7BT
United Kingdom

United States
Robotic Industries Association (RIA)
One SME Drive
P.O. Box 1366
Dearborn, MI 48121
U.S.A.

West Germany
Fraunhofer Institute of Manufacturing
Engineering and automation
Nobelstrasse 12
7000 Stuttgart 80
West Germany

Yugoslavia
Robotics Dept.
Mihailo Pupin Institute
POB 906
11000 Beograd SFR,
Yugoslavia

Robot Manufacturers

Austria
IGM Industrie Zentrum
Mo-sud Strasse
2a Halle M8
A–2351 Wiener Neudorf
Austria

Voest Alpine AG
Postfach 2
A 4010 Linz
Austria (tel: 0732 585 1)

Belgium
Distribel
33 rue Godwin, Ensival
B–4850 Verviers
Belgium (tel: 087 33 11 56)

FN Robotics SA
Shell Building
60 rue Ravenstein
B–1000 Brussels
Belgium (tel: 02 511 2500)

L.V.D.
Nijerheidslaan
2–8630 Gullegem
Belgium

Tecnomatix
Herentalsebaan
71–2100 Deurne
Belgium

Woit & Cotrico
rue de Compas
19–1070 Bruxelles
Belgium

Canada
Can Engineering Sales
P.O. Box 428
6800 Montrose Road
Niagara Falls, Ontario
Canada

Diffracto Ltd.
2775 Kew Drive
Windsor, Ontario
Canada NVT 159

Pavesi International
Burlington
Ontario
Canada (tel: 416 631 6909)

Wexford Robotics Ltd.
2118 Queen
Regina, Sask.
Canada S4T4C3 (tel: 306 522 7469)

Finland
Nokia Robotics AB
Matinkatu 22
02230 Espoo 23
Finland (tel: 358 0 8035610)

Oy Nokia Ab Robotics
Pursimiehenkatu 29-31
SF–00150 Helsinki 15
Finland

Oy W Rosenlaw AB
PB 51
SF–28101 Pori 10
Finland

France
ACMA-Cribier (Renault)
3–5 rue /denis Papin
95250
France (tel: 3/4135490)

Afma Robots
BP 315, St. Avertin
37173 Chambray les Tours Cedex
France (tel: 47/276066)

AKR (Aoip Kremlin Robotique)
6 rue Maryse Bastie
ZI de Sait-Geunault
91031 Evry Cedex
France (tel: 6/077 9615)

Bertin & Cie
B.P. 3
78370 Plaisir
France (tel: 3/056 25 00)

C.G.M.S.
98 rue D'Ambert
B.P. 1825
45008 Orleans Cedex
France (tel: 38/86 25 14)

Citroen
133 quai Andre Citroen
75747 Paris, Cedex 15
France (tel: 578 61 61)

Commercy
55200 Commercy
France (tel: (29) 010104)

Continental Parker
51 rue Pierre
92110 Clichy
France (tel: 793 3330)

Dimenco PSP
16 rue Gay Lussac
25000 Besancon
France (tel: 81/53 81 32)

DOGA
avenue Gutenburg
B.P. 53, 78311 Maurepas Cedex
France (tel: 3/062 41 41)

Durbus
40 rue Marceau
93100 Montreuil
France (tel: 859 51 84)

Duffour et Igon
Rue de l'Oasis
31300 Toulouse
France (tel: 61/42 35 36)

H. Ernault-Somua
32 avenue de l'Europe
78140 Velizy-Villacoublay
France (tel: 946 96 40)

France Euromatic
8 rue du Commerce
68400 Reidsheim
France (tel: 89/64 15 33)

Holbronn Freres
4 rue Jeane Moulin
94130 Nogent sur Marne
France (tel: (1) 873 6945)

Industria
28 avenue Clara
94420 Le Plessis Trevise
France (tel: 01/576 53 78)

Kasto-France
6 rue Pierre et Marie Curie
94430 Chennevieres sur Marne
France (tel: (1) 203 0381)

Lanquepin
89 rue Proudhon
93214 La Plaine St Denis
France (tel: 55/71 44 11)

Pharemme
Les Nouhauts B.P. 1
877370 Saint Sulpice Lauriere
France (tel: 55/71 44 11)

SCEMI
61 rue de fungas
38300 Bourgoin-Jallieu
France (tel: 74/93 20 04)

Sciaky
119 quai Jules Guesde
94400 Vitry/Seine
France (tel: 680 85 07)

Sietam
38–48 avenue du President Kennedy
91170 Vitry-Chatillon
France (tel: 6/996 91 80)

Sodimat
Sapignies 62121 A
62121 Achiet-le Grand, Cedex 6
France (tel: 6/903 78 79)

Sormel SA
rue Becquerel
Z.I. Chateaufarine, B.P. 1565
25009 Besancon Cedex
France (tel: 8181 4245)

Italy
AISA
via Roma 20
26020 Cumigano
Italy

Ansaldo SpA
viale Sarca 336
20126 Milan
Italy

Basfer SpA
via Iseo 60
20052 Monza
Italy

Camel Robot srl
piazza Addolorata 5
20030 Palazzolo Milanese, Milan
Italy

Comau SpA
via Rivalta 30
10095 Grugliasco, Turin
Italy

DEA SpA
Corso Torino 70
10024 Moncaliere, Turin
Italy

Duplomatic SpA
via Alba 18
21052 Busto Arsizio, Varese
Italy

FATA-Bisiach & Carru
Strada Statele 24, 12km
10044 Pianezza, Turin
Italy

Gaitto Impianti SpA
Statele Milano-Crema km 27
26100 Vaiano Cremasco, Cremona
Italy

Jobs SpA
via Marcoloni 11
29100 Piacenza
Italy

Norda SpA
via Vallecamonica 14/F
25100 Brescia
Italy

Olivetti OCN SpA
Stradele Torino
10090 S
Barnado D'Ivrea, Turin
Italy

Prima Progetti SpA
Strada Carigano 48/2
10024 Moncalieri, Turin
Italy

Robox Elettronica Industriale
36 via Sempione-Strada Privata Mainini
28053 Castelletto Ticino Novara
Italy (tel: 0331 922086)

Japan
Aida Engineering Ltd.
Automatic Machine Department
No. 2–10 Oyama-cho, Sagamihara-Shi
Kanagawa-Ken 229
Japan (tel: 0427 (72) 5231)

Citizen Watch Company
840 Shimotomi Takeno Tokorozawa City 359
Saitama Pref.
Japan (tel: (0429) 42-6271)

Daido Steel Company Ltd.
7–13 Nishi Shinbashi 1–chome
Minato-ku, Tokyo
Japan (tel: 03/501 5261)

Daikin Kogyo Company, Ltd.
700–1 Hitotsuya
Settsu City, Osaka 564
Japan (tel: 06/349 7361)

Dainichi Kiko Company Ltd.
Kosiacho Kakakoma-gun
Yamanashi Prefecture
Japan (tel: 05528/2 5581)

Fanuc Ltd.
Engineering Administration Department
5-1 Asahiugaoka 3-chome
Hino City, Tokyo
Japan (0425/84 1111)

Fuji Electric Company Ltd.
12-1 Yuraku-cho 1-chome
Chiyoda-ku, Tokyo 100
Japan (tel: 03/211 7111)

Fujitsu Ltd.
1015 Kamiodanaka Nakahara-ku
Kawasaki-City Kana 211
Gawa-Pref.
Japan (tel: (044) 777-1111)

Harmo Japan
7621-10 Fujizuka
Nisha-minoco, Ina-City
Nagano Pref.
Japan (tel: 399-45)

Hirata Industrial Machines
5-4 Myotaiji-machi
Kumamoto 860
Japan

Hitachi Ltd.
Industrial Components and Equipment Division
4-1 Hammatsu-cho, 2-chome
Minato-ku
Tokyo
Japan (tel: 03/435 4272)

Hikawa Industry Company, Ltd.
22-1 Futamuradai 1-chome
Toyoake-City 470-11
Aichi-Pref.
Japan (tel: (05613) 4-1611)

Ichikoh Engineering Company Ltd.
1297-3 Ninomiuya-cho
Maebashi City
Japan (tel: 2072/68 2131)

Ikegai Iron works
1-21 Shiba 4-chome
Minato-ku
Tokyo
Japan 108 (tel: (03) 452-8111)

Ishikawajima-Harima Heavy
Industries Co., Ltd.
Shin-Ohtemachi Bldg., 2-1, Ohtemachi
2-chome, Chiyoda-ku
Tokyo 100
Japan (tel: (03) 244-6496)

Kayaba Industry Company Ltd.
Engineering Administration Department
4-1 Hammatsu-cho 2-chome
Minato-ku, Tokyo
Japan (tel: 03/435 3511)

Kawasaki Heavy Industries Ltd.
Hydraulic Machinery Div.
4-1 Hammatsu-cho 2-chome
Minato-ku, Tokyo
Japan (tel: 03/435 6853)

Kitamura
1870 Toide-cho
Takaoka City, Toyama Pref., 939-11
Japan (tel: 0766/63-11000)

Kobe Steel Ltd.
Machinery & Engineering Dept.
8-2 Marunouchi 1-chome
Chiyoda-ku, Tokyo
Japan (tel: 03/218 7553)

Komatsu Ltd.
3-6 Akasaka 2-chome
Minato-ku, Tokyo 107
Japan (tel: 03/584 7111)

Koyo Automation Systems Company Ltd.
26-3 Toyocho 1-chome
Koto-ku, Tokyo 135
Japan (tel: 03/615 2611)

Kurogane Crane Company Ltd.
60 Shibacho, Minami-ku
Nagoya City 457
Japan (tel: 052/822 3211)

Kyoritsu Engineering Company, Ltd.
Miyado Building, 6-19 Hacchobori
Naka-ku Hiroshima Pref
Japan (tel: 0822/28 9747)

Kyoshin Electric Company Ltd.
20-7 Ikegami 6-chome
Ota-ku, Tokyo 146
Japan (tel: 03/751 2131)

Marol Company Ltd.
1.34 2-chome, Ohashi-cho
Nagata-ku, Kobe
Japan (tel: (078) 611 2151)

Matsushita Industrial Equipment Company Ltd.
2-7 Matsuba-Cho
Kadoma City, Osaka 571
Japan (tel: 06 901-1171)

Meidensha Electric Mfg. Company Ltd.
Mechatronics Business Division
2-1-17 Osaki Shinagawa-ku
Tokyo
Japan 104 (tel: 492/1111)

Mitsubishi Electric Corporation
Engineering Department
2-3 Marunouchi 2-chome
Chiyoda-ku, Tokyo
Japan 100, (tel: 03/218 2111)

Mitsubishi Heavy Industries Company Ltd.
Precision & Machinery Division
5-1 Marunouchi 2-chome
Chiyoda-ku, Tokyo
Japan 100 (tel: 03/212 3111)

Mizano Iron Works
Kanimachi Kanigun
Gifu Pref. 509 02
Japan

Motoda Electronics Company Ltd.
Kamikitazawa 4-chome
Setagaya-ku, Tokyo 156
Japan (tel: 03/303 8491)

Murati Machinery Ltd.
3 Minamiochiai-cho Kishoin Minami-ku
Kyoto City 100
Japan (tel: (075) 681-9141)

Nachi-Fujikochi Corporation
Machine Tool Division, World Trade Centre
4-1 Hammatsucho, 2-chome, Minato-ku
Tokyo
Japan (tel: 03/435 5111)

Nagoya Kiko
38 Mori Koshi, Shinden-cho
Toyoake City
Aichi Pref.
Japan (tel: 03/451 5131)

Nippon Robot Machine Company Ltd.
73 Yonge Nihongi-cho
Anjo City, Aichi Pref. 446
Japan (tel: 0566/74 1101)

Nitto Seiko Company Ltd.
Umegahata 20, Inokura-cho, Ayabe /city
Kyoto 623
Japan (tel: (0773) 42-3111)

Okamura Corporation
2944 Urazato 5-chome
Yokosuka City, Kanagawa Pref. 237
Japan (tel: 0468/65 8201)

Oki Electric Industries Company Ltd.
7-12, Toranomon 1-chome, Minato-ku,
Tokyo 125
Japan (tel: (03) 501-3111)

Osaki Denki Company Ltd.
3-31, 4-chome, Nishimikuni Yodogawaku
Osaka 532
Japan (tel: (06) 394-1191)

ORII Corporation
6 Suzuhawa, Isehara City,
Kanagawa Pref. 259-11
Japan (tel: 0463 93-0811)

Osaka Transformer Company Ltd.
1-11 Tagawa 2-chome
Yodogawa-ku, Osaka 532
Japan (tel: 06/301 1212)

Pental Company Ltd.
1-8 Yoshi-cho, 4-chome
Soka City
Saitoma Pref. 340
Japan (tel: 0489 22:1111)

Sanki Engineering Company Ltd.
Sanshin Building
4-1 Yurakucho 1-chome Chiyoda-ku, Tokyo
Japan 100 (tel: 03/502 6111)

Sankyo Seiki Manufacturing Company Ltd.
17-2 Shinbashi 1-chome
Minato-ku, Tokyo
Japan 105 (tel: 03/508 1156)

Shawa Kuatuski
3-19 Kanda-Sakumacho
Chiyada-ku, Tokyo
Japan

Shinko Electric Company Ltd.
3-12 2 Nihonbashi
Chuo-ku, Tokyo 103
Japan (tel: 03/274 1111)

Shinmeiwa Industry Company Ltd.
1-1 Shinmeiwa-cho
Takarazuka City, Hyogo Pref.
Japan 665 (tel: 0798/52 1234)

Shoku Corporation
1010 Minorudai, Matsudo City
Chiba Pref.
Japan (tel: 0473/64 1211)

Star Seiki Company Ltd.
252 Kawachiya Shirden
Komaki City, Aichi Pref. 485
Japan (tel: 0568/75 5211)

Sumitomo Heavy Industries Ltd.
Shumisho Building, 1 Mitoshirocho
Chiyoda-ku, Tokyo
Japan (tel: 03/296 5183)

Taiyo Ltd.
48 Kitaguchi-cho
Higashiyodogawa-ku
Osaka
Japan 533 (tel: 06/340 1111)

The Japan Steel Works Ltd.
Hibiya Mitsui Bldg. 1–2 Yurakucho 1–chome
Chiyoda-ku, Tokyo 100
Japan (tel: (03) 501-6111)

Tokico Ltd.
6–10 Uchikanda 1–chome
Chiyoda-ku, Tokyo
Japan 101 (tel: 03/292 8111)

Tokyo Keiki Company Ltd.
2–16 Minami Kamata
Ohta-ku, Tokyo 144
Japan (tel: 03/732 2111)

Tokyo Shibaura Electric Company Ltd.
1–6, Uchi-Saiwaich 1–chome
Chiyoda-ku, Tokyo 144
Japan

Toshiba Seiki Company Ltd.
14–33 Higashi Kashiwagaya 5–chome
Ebina City, Kanagawa Pref.
Japan 243 (tel: 0462/31 8111)

Toyota Machine Works Ltd.
1 Asahi-cho, 1–chome
Karia City Aichi Pref. 448
Japan (tel: 0566/22 2211)

Yaskawa Electric Manufacturing Company Ltd.
Chiyoda-ku
6–1 Ohtemachi 1–chome
Tokyo
Japan 100 (tel: 03/217 4111)

Yasui Sangyo Company Ltd.
3711 Mannohara-Shinden
Fujinomiya-Shi
Shizuoka-ken
Japan (tel: 05442 62124)

Norway
Oglaend
4301 Sandes P.B. 115
Norway (tel: 04 605000)

Trallfa
P.O. Box 113
N 4341 Byrne
Norway (tel: 04 48 1800)

Spain
Campania Anomina de Electrodos
Infanta Carlota 56
Barcelona
Spain (tel: (93) 666 50152)

Inser SA
Jose Ortega y Gasset 62
Madrid, Spain

Oficina Technia Comercial (OTC)
Padilla, 382.5
Barcelona
Spain (tel: (93) 309 6462)

Sweden
ASEA AB
Industrial Robot Division
S 72183 Vasteras
Sweden (tel: 021 100000)

ASEA AB (previously Electrolux)
Industrial Robot Division Stockholm
Fagelviksvagen 3
S–145 53 Norsborg
Sweden (tel: 046 753 89100)

Atlas Copco
S–105 23
Stockholm, Sweden

ESAB AB
Box 8004
S–40277 Gothenburg
Sweden

Statt-Kaufeldt AB
P.O. Box 32 006
S–12611 Stockholm
Sweden (tel: 08 810100)

Spine Robotics AB
Flojelbergsgaten 14
S 43137 Molindal
Sweden (tel: 031 870710)

Torsteknik AB
Box 130
S–385 00 Torsas
Sweden

Switzerland
Automelec S.A.
Case postale 8
137, rue des Pondireres
CH-2006 Neuchatel
Switzerland

Cod Inter Techniques S.A.
16, rue Albert-Gos
CH-1206 Geneve
Switzerland

Ebosa Kapellstrasse 26
CH 2540 Grenchen
Switzerland

Microbo
3 avenue Beauregard
CH 2035 Corcelles
Switzerland (tel: 1941 3831 5731)

Schweissindustrie Oerlikon Buhrle AG
Birchstrasse 230
8050 Zurich
Switzerland (tel: 01 301 2121)

United Kingdom
Airstead Industrial Systems Ltd.
New England House
New England Street
Brighton BN1 4GH
UK (tel: 0273 689793)

Ajax Machine Tool Ltd.
Knighton Heath Estate
847/855 Ringwood Road
Bournemouth Bh1
UK

ATM Engineering Ltd.
Unit 9, Earls Way
Church Hill Ind. Est.
Thurmaston, Leicester LE4 8DH
UK (tel: 0533 693396/7)

British Federal Welder and Machine Co. Ltd.
Castle Mill Works
Birmingham New Road
Dudley, West Midlands DY1 4DA
UK (tel: 0384 54701)

Cirrus Equipment Ltd.
Heming Road
Redditch
Worcs B98 0DN
UK (tel: 0527 27882)

Fairy Automation Ltd.
Techno Trading Estate
Bramble Road, Swindon
Wilts SN2 6HB
UK (tel: 0793 481161)

Frazer Nash
Vine House
143 London Road
Kingston-upon-Thames
Surrey KT2 6NW
UK

H.H. Freudenberg Automation
Cobden House, Cobden Street
Leicester, LE1 2LB
UK

GEC Robot Systems Ltd.
Boughton Road
Rugby CV21 1BD
UK (tel: 0788 2144)

George Kuikka Ltd.
Hill Farm Avenue
Leavesden, Watford
Herts, UK (tel: 09273 70611)

Haynes & Fordham
Unit 4, Moorfield Ind Est
Yeadon
Leeds LS19 7BM
UK (tel: 0532 507090)

INA Automation Ltd.
Forge Lane, Minworth
Sutton Coldfield
West Midlands B76 8AP
UK (tel: 021 351 4047)

Lamberton Robotics Ltd.
26 Gartsherrie Road
Coatbridge
Strathclyde ML5 2DL
UK (tel: 0236 26177)

Lansing Industrial Robots
Kingsclere Road
Basingstoke
Hants
UK (tel: 0256 3131)

Lincoln Electric Ltd.
Welwyn Garden City
Herts AL7 1QA
UK (tel: 070732 24581)

Martonair Ltd.
St. Margarets Road
Twickenham TW1 1RJ
UK (tel: 01 892 4411)

Marwin Production Machines
Waddons Brook
Wednesfield
Wolverhampton WV11 3AA
UK (tel: 0902 65363)

Modular Robotic Systems Ltd.
30/31 St. George's Square
Worcester WR1 1HX
UK (tel: 0905 612881)

Pendar Robotics Ltd.
Unit 10, Rassau Industrial Estate
Ebbw Vale
Gwent NP3 5SD
UK (tel: 0495 307070)

Remek Micro Electronics
35 Barton Road, Water Eaton
Industrial Estate
Bletchley, Milton Keynes MK2 3HY
UK

Ringway Power Systems Ltd.
Churchill House, Talbot Road
Old Trafford
Manchester M16 0PD
UK (tel: 061 872 6829)

Taylor Hitec Ltd.
77 Lyons Lane
Chorley, Lancs PR6 0PB
UK (tel: 02572 65825)

Wickman Automated Assembly Ltd.
Herald Way, Brandon Road
Binley
Coventry CV3 2NY
UK (tel: 0203 45080)

WRA Ltd.
Untis 2/3, Wulfrun Trading Estate
Stafford Road
Wolverhampton
UK (tel: 0902 711201)

United States
Accumatic Machinery Corp.
3537 Hill Avenue
Toledo, Ohio 43607
(tel: 419-535-7997)

Acrobe Positioning Systems Inc.
3219 Dolittle Drive
Northbrook, Illinois 60062

Action Machinery Company
P.O. Box 3068
Portland, Oregon 97208

Admiral Equipment Company Ltd.
305 West North Street
Akron, Ohio 44303
(tel: 216-253-1353)

Advanced Robotics Corporation
Route 79
Newark Industrial Park
Hebron, Ohio 43025
(tel: 614-929-1065)

Ameco Corporation
P.O. Box 385
Menomonee, Wisconsin 53051

American Robot Corporation
201 Miller Street
Winston-Salem, North Carolina 27103
(tel: 919-748-8761)

Anorad
110 Oser Avenue
Hauppauge, New York 11788
(tel: 516-231-1990)

Armax Robotics Inc.
38700 Grand River Avenue
Farmington Hills, Michigan 48018
(tel: 313-528-3630)

ASEA Inc.
1176 E. Big Beaver
Troy, Michigan 48084
(tel: 313-528-3630)

Automatix Inc.
1000 Tech Park Drive
Billerica, Massachusetts 01821
(tel: 617-667-7900)

Barrington Automation Ltd.
1002 South Road
Fox River Grove, Illinois 60021

Binks Corporation
9201 West Belmont Avenue
Franklin Park, Illinois 60131
(tel: 312-671-3000)

Ceeris International Inc.
1055 Thomas Jefferson St. NW
Suite 414
Washington D.C. 20007
(tel: 202-342-5400)

Cincinnati Milacron
Industrial Robot Division
215 S. West Street
Lebanon, Ohio 45036
(tel: 513-932-4400)

Comet Welding Systems
900 Nicholas Road
Elk Grove Village, Illinois 60007
(tel 312-956•0126)

Control Automation Inc.
P.O. Box 2304
Princeton, New Jersey 08540

Cybotech Corporation
P.O. Box. 88514
Indianapolis, Indiana 46208
(tel: 317-298-5136)

Cyclomatic Inc.
7520 Corvey Court
San Diego, California 92111
(tel: 619-292-7440)

Dependable-Fordath Inc.
400 SE Willimette Street
Sherwood, Oregon 97140

DeVilbiss Company
837 Airport Boulevard
Ann Arbor, Michigan 48104
(tel: 313-668-6765)

Dynamcac Inc.
410 Forest Street
Marlboro, Massachusetts 01752

Everett/Charles
Automation Modules Inc.
9645 Arrow Route, Suite A
Rancho Cucamonga, California 91730
(tel: 714-980-1525)

EWAB America
292 Palatine Road
Wheeling, Illinois 60090

Fared Robotic Systems Inc.
3860 Revere Street, Suite D
P.O. Box 39268
Denver, Colorado 80239
(tel: 303-371-5868)

Fleximation Systems Corporation
53 Second Avenue
Burlington, Massachusetts 01803

GCA/PAR
3460 Lexington Avenue North
St. Paul, Minnesota 55112
(tel: 612-484-7261)

General Electric
Automation Systems
1285 Boston Avenue
Bridgeport, Connecticut 06602
(tel: 203-382-2876)

General Numeric Corporation
390 Kent Avenue
Elk Grove Village, Illinois 60007
(tel: 312-640-1595)

GMF Robotics Corporation
5600 New King Street
Troy, Michigan 48098

Graco Robotics Inc.
12899 Westmore Avenue
Livonia, Michigan 48150
(tel: 313-261-3270)

Hellstar Corporation
1600 N. Chestnut
Wahoo, Nebraska 68066

Hobart Brothers Co.
600 West Main Street
Troy, Ohio 45373

Hodges Robotics International Corporation
3710 North Grand River Avenue
Lansing, Michigan 48906
(tel: 517-323-7427)

IBM Advanced Manufacturing Systems
1000 NW 51st Street
Boca Raton, Florida 33432
(tel: 305-998-2000)

Industrial Automates Inc.
6123 W. Mitchell Street
Milwaukee, Wisconsin 53214
(tel: 414-327-5656)

Intarm
P.O. Box 53
Dayton, Ohio 45409
(tel: 518-294-0834)

Intelledex Inc.
33840 Eastgate Circle
Corvallis, Oregon 97333

International Robotmation Intelligence
2281 Las Palmas Drive
Carlsbad, California 92008
(tel: 619-438-4424)

ISI Manufacturing Inc.
31915 Groesbeck Highway
Fraser, Michigan 48026
(tel: 313-294-9500)

Keller Technology Corporation
Robotics Automation Systems
2320 Military Road
Tonawanda, New York 14150

Lamson Corporation
P.O. Box 4857
Syracuse, New York 13221
(tel: 315-432-5467)

Livernois Automation Co.
25315 Kean
Dearborn, Michigan 48124
(tel: 312-278-0201)

Lynch Machinery Corp.
2300 Crystal Street
P.O. Box 2477
Anderson, Indiana 46018
(tel: 317-643-6671)

Machine Intelligence Corporation
330 Potrero Avenue
Sunnyvale, California 94086
(tel: 408-737-7960)

Mack Corporation
3695 East Industrial Drive
Flagstaff, Arizona 86001
(tel: 602-526-1120)

Manca Inc.
Link Drive
Rockleigh, New Jersey 07647
(tel: 201-767-7227)

Microbot Inc.
453-H Ravendale Drive
Mountain View, California 94943
(tel: 415-968-8911)

Mobot Corporation
980 Buenos Avenue
San Diego, California 92110
(tel: 619-275-4300)

Pickomatic Systems Inc.
37950 Commerce Drive
Sterling Heights, Michigan 48077
(tel: 313-939-9320)

Positech Corporation
114 Rush Lake Road
Laurens, Iowa 50554

Prab Robots Inc.
5944 E. Kilgore Road
Kalamazoo, Michigan 49003
(tel: 616-862-1124)

Precision Robots Inc.
6 Carmel Circle
Lexington, Massachusetts 02173
(tel: 617-862-1124)

Reeves Robotics
Box S.
Issaquah, Washington 98027
(tel: 206-392-1447)

Ron-Con Ltd./Bra-Con Industries
12001 Globe Road
Livonia, Michigan 48154

Robotic Sciences International Inc.
2709 South Halladay
Santa Ana, California 92705
(tel: 714-979-6831)

Robotiks Inc.
507 Prudential Road
Horsham, Pennsylvania 19044
(tel: 215-674-2800)

Sandhu Rhino Robots
308 S. State Street
Champaign, Illinois 61820
(tel: 217-352-8485)

Schrader-Bellows
US Rt. 1 N.
Wake Forest, North Carolina 27587
(tel: 919-556-4031)

Seiko Instruments USA Inc.
2990 W. Lomita Boulevard
Torrance, California 90505
(tel: 213-530-8777)

Sigma Sales Inc.
6505C Serrano Avenue
Anaheim Hills, California 92807
(tel: 714-974-0166)

Sterling Detroit Company
261 E. Goldengate
Detroit, Michigan 48203
(tel: 313-366-3500)

Sterlich
P.O. Box 23421
Milwaukee, Wisconsin 53223
(tel: 414-354-0493)

TecQuipment Inc.
P.O. Box 1074
Acton, Massachusetts 01720
(tel: 617-263-1767)

Thermwood Corporation
P.O. Box 436
Dale, Indiana 47523
(tel: 812-937-4476)

Unimation Inc.
Shelter Rock Lane
Danbury, Connecticut 06810
(tel: 203-744-1800)

United States Robots Inc.
1000 Conshohocken Road
Conshohocken, Pennsylvania 19428

United Technologies Automotive Group
5200 Auto Club Drive
Dearborn, Michigan 48126

VSI Automation Assembly Inc.
165 Park Street
Troy, Michigan 48084

Westinghouse Electric Corporation
(also see Unimation)
400 Media Drive
Pittsburgh, Pennsylvania 15205
(tel: 412-778-4347)

West Germany
Carl Cloos Schweisstechnik GmbH
D-6342 Haiger
West Germany (tel: 02773850)

Fibro GmbH
Postfach 1120
D-6954 Hassmersheim
West Germany

Gebr. Felss
7535 Konigsbach Stein 2
Gutensbergstr. 4
West Germany

G.D.A.
5 Am Bahnhof
D-8915 Fuchstal
West Germany (tel: 08243 2012)

Jungheinrich KG
Friedrich-Ebert-Damm 184
2000 Hamburg 70
West Germany (tel: 040 66 43 50)

KUKA Schweissanlangen + Roboter
Zugspitzstrasse 140
D-8900 Augsburg 43
West Germany (tel: 0821 7971)

Mantec GmbH
Postfach 2620
D-8520 Erklangen
West Germany (tel: 09131/16200)

Messer Griesheim GmbH
Landsbergerstrasse 432
D-8000 Munich 60
West Germany

Nimak-MAG
Postfach 192
D-5248 Wissen
West Germany (tel: 02742 4024/4025)

Ottensener Eisenwerk GmbH
Steinwerder
D-2000 Hamburg 11
West Germany (tel:040 306859)

Pfaff
Postfach 3020/3040
D-6750 Kaiserslautern
West Germany (tel: 0631 881)

Produtec GmbH
Heilbronnerstrasse 67
D-7000
Stuttgart 1
West Germany

Robert Bosch GmbH
Geschaftsbereich Industrieausrustung
7000 Stuttgart 30, Kruppstrasse 1
Postfach 300220
West Germany (tel: 0711 811 5225)

Siemens AG
Rupert-Mayer-Strasse 44
8000 Munchen 70, Postfach 70 00 75
West Germany (tel: 098 722 26126)

Union Carbide Deutschland GmbH
(also see Nimak-MAG)
Postfach 133, D-5248 Wissen
West Germany (tel: 02742 751)

Index